David J. Murphy

Compass Land Surveying

A Guide For
surveyors, attorneys, landowners and others

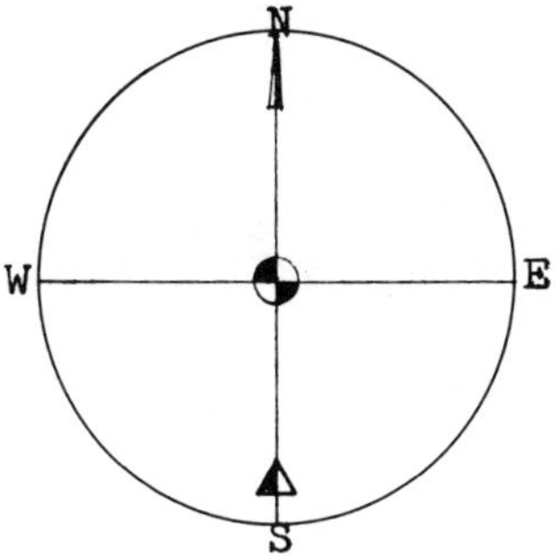

by
F. Henry Sipe

West Virginia Licensed Land Surveyor
West Virginia Registered Professional Forester
Member: West Virginia Association of Land Surveyors
Member: American Congress on Surveying and Mapping
Member: Society of American Foresters

McCLAIN PRINTING COMPANY
PARSONS, WEST VIRGINIA 26287
1974

DEDICATION

This book is dedicated to better land surveying everywhere.

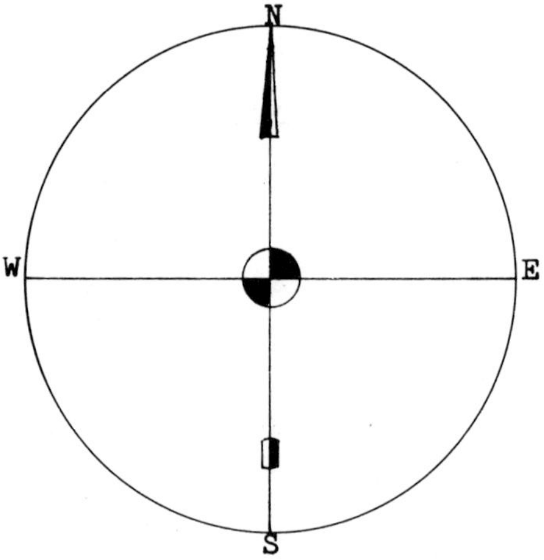

Standard Book Number 87012-084-0
Library of Congress Card Number 75-119818
Printed in the United States of America

1404 Harrison Avenue, Elkins, West Virginia 26241

TABLE OF CONTENTS

PREFACE

About the Book

Surveying land with a compass is neither well-understood nor adequately practiced in many parts of the United States. To improve this condition in West Virginia, the author published a 75-page paperback booklet in 1965 entitled *Surveying Rural Property Boundary Lines in West Virginia.*

Surveyors in other states had similar problems and encouraged the author to expand the subject matter to make it more widely applicable. The second edition in 1970 was a hardback book of 191 pages entitled *Compass Land Surveying.* The 19 chapters of the 1965 booklet were included in the 1970 edition without change.

This third edition is a "complete overhaul". Its main purpose is still to provide simple and "how to do it" instructions for surveying boundary lines of rural or lower value land where fourth-order accuracy is acceptable. The reader will also find in it some aspects of more precise compass and transit surveying. A temptation to shorten the title to *Land Surveying* was suppressed. The book attempts to revitalize the practice of compass land surveying, to show its value and application, and to elevate it to a professional level never before actually attained in this country.

Material inconsistent with the use of the newly developed Warren-Knight open-sight surveyor's compass has been deleted. Results of two years of testing and field performance of the new compass are described. A chapter on most aspects of a misunderstood subject—magnetic disturbance or variation—is included. The chapters on plats, descriptions and other records enlarge and upgrade similar subjects in the 1970 book. A rating form for plats and descriptions is introduced. Miscellaneous items have been added at the end of several chapters.

The book should help not only land surveyors, but prospective surveyors, survey helpers, attorneys, landowners, foresters, realtors, bankers, public officials and others who deal with problems of real estate. Several colleges use it as a textbook. It does not profess to be the "last word" on land surveying; the author learns something from every new job.

Land Surveying—Art or Science?

Dictionaries say "art" is skill, dexterity or power of performing certain actions acquired by experience, study or observation; in one word—"knack". They define "science" as a body of systematized knowledge and facts. In other words, science provides the knowledge and art applies it.

Curtis M. Brown writes that land surveying is based on law, which is not an exact science. Davis and Foote say surveying is the "art of measuring". Clark likens the art of surveying to the medical or legal professions—primarily one requiring mental ability. Brown and Eldridge write that it means the application of knowledge in the mathematical, physical and technical arts and sciences. Russell C. Brinker defines surveying as a science or art. LeRoy E. Lewis, a United States Forest Service surveyor, describes the work of a land surveyor as "the ability to sift through a maze of conflicting, overlapping and often spurious data, and acting in the role of a quasijudicial figure, arrive at a solution that fairly and impartially protects the interests of all parties concerned with that property line".

There are four basic fields involved: law, measurements, investigations and reports.

The North Carolina Manual of Practice states that land surveying does *not* include engineering surveys such as for construction of buildings, railways, highways, airports, utility lines, bridges and the like; although land surveying may be needed in those projects when boundary lines, maps and descriptions are involved.

Acknowledgements

The author gratefully acknowledges the help of persons who reviewed parts of the rough draft for the second edition and suggested improvements: L. Bruce Sumner, cadastral surveyor for the United States Forest Service, San Francisco; the late Gerald R. Hyde, Bedford, New Hampshire and Professor Wilfred H. Baker, formerly of the Civil Engineering Department of West Virginia University at Morgantown. Special thanks go to Sumner who authored two chapters. Appreciation is expressed to Galtjo L. Geertsema of Martinsburg, West Virginia, who reviewed the second book and suggested improvements; to Richard W. Kuberry, Chief of the Fredericksburg Geomagnetic Center for comments on magnetic disturbances; to Darrell R. Dean, Jr., and Willis J. Everman whose contributions are found in the book; to Everman and the J. M. Huber Corporation for the "Washington" survey plat; to George Nietzold, A. M. McClain and Dr. David K. Allen for suggestions; to Richard M. Feaster of Elkins for drawings in chapters 15, 23, 24, and 36 and for one plat, and to James B. Dixon for one plat. Thanks also go to Ken McClain and George Smith of the McClain Printing Company for publishing advice, and to my wife Catharine who helped proofread and withstood the strain.

Abbreviations

Common symbols and abbreviations such as #, % and ½ are used in this book, also N for north, NE for northeast, etc. References to books listed in Chapter 37 are sometimes shortened and italicized. Where appropriate, the following additional abbreviations are used at times:

Article	Art
chain(s)	ch
Chapter	Chap
clockwise	cw
counterclockwise	ccw
cosine	cos
Deed Book	DB
degree	°
departure(s)	Dep(s)
diameter breast height	dbh
distance	dist
double meridian distance	DMD
elongation	E
equals, is equal to	=
foot, feet	ft
Forest Service	FS
Government	Govt
Govt. Printing Office	GPO
Horizontal	Hor
inch	"
iron pipe	IP
latitude(s)	Lat(s)
link(s)	l
longitude	Long
minute(s) of arc	'
multiply(ied) by	x
observation point	OP
page	p
pole	p
right of way	r/w
sine	sin
square	sq
United States	U. S.
West Virginia	W. Va.

About the Writer

The author is a consulting land surveyor and professional forester. He was graduated in 1925 from the Pennsylvania State Forest School at Mont Alto, Pennsylvania, with degree of Bachelor of Science in Forestry. In 1929 he was graduated from Yale University School of Forestry with degree of Master of Forestry. In college he received education and training in surveying.

Since 1929 he worked for three years for the States of Pennsylvania and North Carolina, and 32½ years for the Forest Service, U. S. Department of Agriculture. During these 35½ years as a forester, he performed survey work as part of his duties. Since retirement from the Forest Service in January, 1962, he has engaged in land surveying for private clients.

In 1968 he organized the West Virginia Association of Land Surveyors and was elected its first President. In 1969 he assisted in obtaining passage of a law to require land surveyors to be licensed in West Virginia and was appointed to its first board of examiners.

AMERICAN CONGRESS ON SURVEYING AND MAPPING ON MARCH 7, 1970 ADOPTED THE FOLLOWING POLICY ON LAND SURVEYING:

1. Land Surveying in the United States is a separate and distinct field of professional practice, based on a unique educational background. It requires entirely separate licensing and fully qualified representation on examining and licensing boards.

2. The complex of licensing laws and regulations within each state should assure that those licensed to practice in each of the professions are educated, experienced and competent to practice the profession, as a means for safeguarding the public interest.

Chapter 1

HISTORY OF SURVEYS AND LAND TITLES

Background

The American Indians did not have property laws as we know them. Certain tribes laid claim to general areas, while places like Kentucky were hunting grounds for nomadic tribes. English common law provided the basis for law in the 13 original colonies, but Statute and Case law have largely replaced it.

Before the Revolutionary War the King of England made grants of land in America. Later the states assumed ownership of all land and issued land grants. In 1784 all unallocated land in Virginia became state land, and until 1863 West Virginia was a part of Virginia. George Washington and Thomas Jefferson were early surveyors. Washington made surveys and owned land in West Virginia (see next page).

It was the policy in Virginia—and most other states—to dispose of its land to encourage its settlement and development, and to obtain taxes to run its government. In 1779 Virginia established a Land Office to grant "waste and unappropriated land". Warrants were issued upon payment of a fee; these allowed the warrantee to locate land, have it surveyed, and return the survey to the registrar within a fixed time; the Governor then issued a grant (patent). Grants could cover several warrants or parts of warrants, which were often assigned to other people even before survey. Grants are on record in the State Land Office in Richmond, Virginia; copies of some are found in courthouses in West Virginia, and in the State Auditor's office in the Capitol at Charleston.

Most of the land east of the Appalachian Mountains was granted by 1800; to the west, after 1800. Many early grants were issued to speculators who never saw the land, did not intend to develop it, and did not pay taxes. "Blanket" grants were issued also; they included all the land in a defined area except what had been previously patented. Field surveys were often poor; perhaps the closing line was not run. Monuments were not set, even if the grant said they were; corners were not marked. Surveys and grants overlapped. Because taxes were needed to run the government, the taxation of land was closely related to the problems of surveys and titles. The Virginia Legislature tried to correct these inequities without complete success.

Early surveys were made by compass and magnetic north. Straight lines were sometimes run by the picket method—lining up stakes in a straight line. Distances were first measured with ropes, sticks, etc. These were gradually replaced by a chain invented by Edmond Gunter; it was 66 feet long and consisted of 100 wire links connected by interlocking

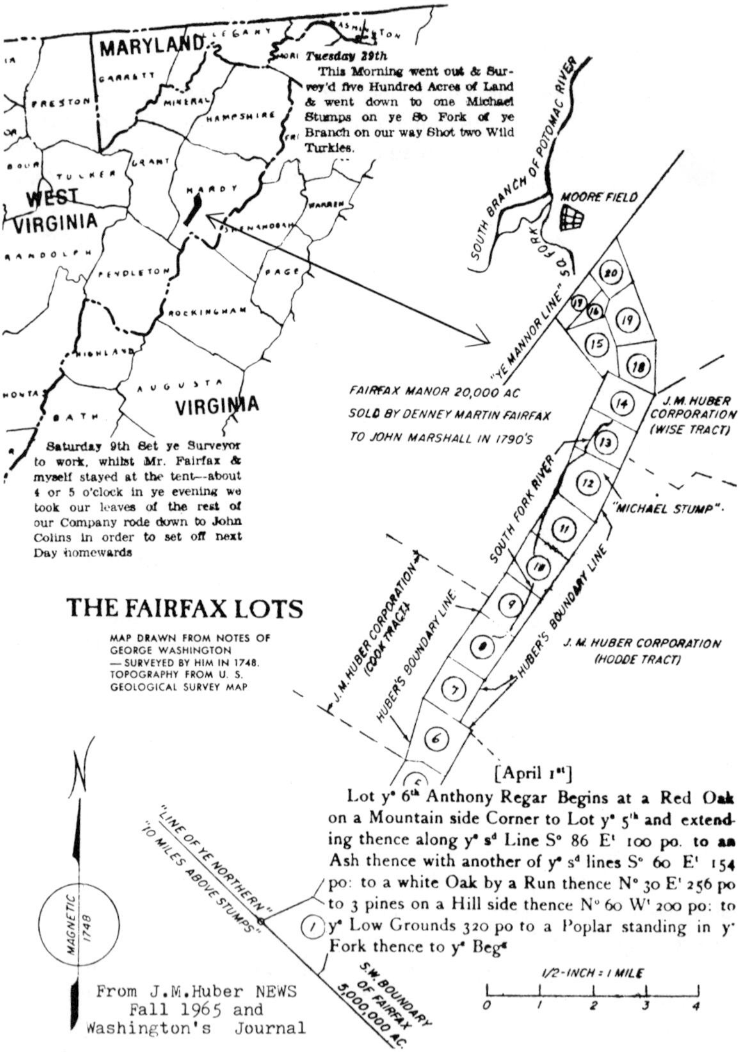
MARYLAND
LEGANY
WASHINGTON
GARRETT
PRESTON
MINERAL
HAMPSHIRE
TUCKER
GRANT
HARDY
WEST VIRGINIA
WARREN
SHENANDOAH
RANDOLPH
PENDLETON
PAGE
ROCKINGHAM
HIGHLAND
AUGUSTA
BATH
VIRGINIA
Tuesday 29th
This Morning went out & Survey'd five Hundred Acres of Land & went down to one Michael Stumps on ye So Fork of ye Branch on our way Shot two Wild Turkies.
Saturday 9th Set ye Surveyor to work, whilst Mr. Fairfax & myself stayed at the tent—about 4 or 5 o'clock in ye evening we took our leaves of the rest of our Company rode down to John Colins in order to set off next Day homewards
SOUTH BRANCH OF POTOMAC RIVER
MOORE FIELD
SO. FORK
"YE MANNOR LINE"
FAIRFAX MANOR 20,000 AC
SOLD BY DENNEY MARTIN FAIRFAX
TO JOHN MARSHALL IN 1790'S
J. M. HUBER CORPORATION (WISE TRACT)
"MICHAEL STUMP"
SOUTH FORK RIVER
HUBER'S BOUNDARY LINE
J. M. HUBER CORPORATION (COOK TRACT)
J. M. HUBER CORPORATION (HODDE TRACT)
THE FAIRFAX LOTS
MAP DRAWN FROM NOTES OF GEORGE WASHINGTON — SURVEYED BY HIM IN 1748. TOPOGRAPHY FROM U. S. GEOLOGICAL SURVEY MAP
N
MAGNETIC 1748
"LINE OF YE NORTHERN"
"10 MILES ABOVE STUMPS"
S.W. BOUNDARY OF FAIRFAX 5,000,000 AC.
[April 1st]
Lot ye 6th Anthony Regar Begins at a Red Oak on a Mountain side Corner to Lot ye 5th and extending thence along ye sd Line So 86 Et 100 po. to an Ash thence with another of ye sd lines So 60 Et 154 po: to a white Oak by a Run thence No 30 Et 256 po to 3 pines on a Hill side thence No 60 Wt 200 po: to ye Low Grounds 320 po to a Poplar standing in ye Fork thence to ye Begn
1/2-INCH = 1 MILE
0 1 2 3 4
From J.M.Huber NEWS
Fall 1965 and
Washington's Journal

loops. 80 chains = 1 mile and 10 square chains = 1 acre. A link is 7.92″ long. A pole (or rod or perch) = 16-1/2 feet or 1/4 chain. The Gunter's chain was gradually replaced by the modern steel tape. In the last 75 years the transit has been used increasingly for surveys of higher value land.

West Virginia was made a state in 1863. The Constitution of 1872 provided that all private land rights derived under the laws of Virginia and all rights established between 1863 and 1872 should remain valid and secure. It provided for handling tax delinquent lands to try to keep them in private ownership. Rural boundary lines in this state are usually described by "metes and bounds" and not by the rectangular system common outside the 13 original colonies.

The Present Situation–The Problem

Under a system that contained many overlapping surveys and poorly marked corners and lines, many disputes arose. Original grants, though hardly locatable on the ground, were later subdivided, often describing corners where none existed or which had been destroyed. Other problems were created by surveying by "surface measure".

Errors in older surveys may have been caused by failure to measure horizontally or by using a magnetic compass. But there is evidence that distances *were* often measured horizontally by holding short tapes–or short sections of a 66 foot chain–level. It is far more likely that these so-called errors were the result of inability of later surveyors to find old corners and lines that were poorly marked.

Some rural property lines in this state have not been resurveyed for many years. Corners and line marks have disappeared; trees have been cut or blown down. Older residents who knew where corners were, have died without passing on the information. Adjoiner corners may be differently described although intended to be the same. Fortunately some lines are marked by roads, streams, ridges and fences.

Poor surveys cannot be laid entirely at the feet of surveyors. Landowners were often unwilling or unable to pay for a good survey. Some just "stepped off" distances for deeds. Many early surveys were made reasonably well; present troubles developed by passage of time without remarking or resurveys. Deed calls are still prepared without surveyor advice and are difficult to survey on the ground.

A start has been made in upgrading the survey of land in West Virginia. The 1969 law as amended, and the rules and regulations have placed a "floor" under the practice of land surveying. As surveyors raise their standards above this floor, progress will be made.

A novel development is the work of Galtjo L. Geertsema, LLS of Martinsburg, West Virginia. Familiar with the Cadastre system of Europe, Geertsema is developing a similar system for counties in his area. He starts by mapping original grants on a 7½ minute quadrangle. Grants are numbered and indexed and the date of survey shown. Subsequent land records emanating from the grants are catalogued. Geertsema's work is for sale to public and private parties.

The Metric System of Measurement

The United States will likely soon change to the metric system of measurements. A few basic converting factors are shown below.

1 foot	=	0.30480	meter
1 rod	=	5.0292	meters
1 chain	=	20.1168	meters
1 square foot	=	0.09290	square meter
1 square rod	=	25.2929	square meters
1 acre	=	4046.86	square meters
1 acre	=	0.404686	hectare
1 square mile	=	258.999	hectares
1 meter	=	3.28084	feet
1 square meter	=	1.19599	square yards
1 hectare	=	2.47105	acres

The First Licensed Surveyors?

The following was one of several laws made at the second session of the General Court held at Boston on October 11, 1682:

"This Court doth Order, that no Countrey Grants of Land shall henceforth be laid out, but by some known able, and approved Person whom this Court shall appoint, who shall upon Oath to Act faithfully in that work, and make Return under their hands of all such Land laid out by them, with the Quantity, in a fair Draught or Plat, together with the Quality as near as they can: In pursuance of this Order. This Court appoints and allows of Capt. Elisha Hutchinson, Mr. Jonathan Danforth, Mr. John Flint, Mr. Samuel Andrews, Mr. David Fish, Mr. William Carr, Mr. Joseph Hawley, and Mr. Samuel Marshfield to be Surveyors, and any one of them to lay out such Countrey Lands as any Person have had, or shall have Grants from this Court. And the said Persons shall take this following Oath before they act in Surveying any such land."

"You swear by the Great Name of the Everliving God, that you will faithfully and impartially to the best of your skill, Survey and Measure such Grants of Land that are made by the General Court to any Person or Persons within this Jurisdiction; and present a fair Draught and Plat of such Land to this Court; therewith certifying the Quality of such Land, all which you shall make Return of to this Court under your hands, so help you God." From the Massachusetts Surveyor and West Virginia Surveyor.

Chapter 2

LEGAL ASPECTS OF BOUNDARY SURVEYS

All surveys must be conducted in accord with the law. The term "resurvey" usually means reestablishing corners and lines of a prior deed, *at the location of the prior survey.* Improper surveys may "give" property to a person who does not rightfully own it, and they are subject to being upset by proper surveys and court decisions. A surveyor must know *why* he does what he does. Virginia laws—both statute and case law—apply in West Virginia unless they have been superseded. There are not many West Virginia statutes on land surveys.

West Virginia case law is more voluminous. Decisions handed down by the State Supreme Court furnish guide lines for the general rules of surveying, particularly as to the weight to be given to various kinds of title and survey evidence. However, case law does not tell the surveyor what techniques to use. This book tries to bridge the gap between the law and the survey.

When all corners or boundaries cannot be retraced or relocated with certainty, the greatest weight must be given in the resurvey to those calls which are most likely to be correct and which therefore take precedence over other calls. The basic rule for retracing former surveys is: *the new surveyor must follow in the footsteps of the original or former surveyor,* often regardless of the mistakes of the former surveyor. He must not correct previous mistakes, nor make a survey where he thinks it should have been made but was not. Work should always be on the safe side. These parts of the Constitution and 1965 Statutes of West Virginia relate to surveying:

Constitution

Art. II, Sec. 1. The State of West Virginia includes the bed, bank, and shores of the Ohio River and so much of the Big Sandy as was formerly included in the Commonwealth of Virginia; and all territorial rights and property in, and jurisdiction over same, heretofore reserved by, and vested in, the Commonwealth of Virginia, are vested in and shall hereafter be exercised by, the State of West Virginia. And such parts of the said beds, banks, and shores as lie opposite, and adjoining the several counties of this State, shall form parts of said several counties.

Art. VIII, Sec. 3. Gives the Supreme Court of Appeals jurisdiction in land title disputes.

Art. IX, Sec. 1. Provides for election of County Surveyor of Lands;

term 4 years. Sec. 4. The County Surveyor may be indicted for malfeasance.

Art. XIII, Sec. 1. Private land rights obtained under Virginia laws, or West Virginia laws enacted before the Constitution was adopted, shall be valid.

Art. XIII, Sec. 2. Prohibits land entries by warrant.

Art. XIII, Sec. 3. Provides for disposal of lands forfeited to the State. Also discusses relation of 10 years actual possession to forfeiture.

Statutes

The following references are to the West Virginia Code as amended through the 1965 Acts. The first number is the Chapter, the second is the Article, and the third is the Section.

6-2-10. Sets bond for County Surveyor at $1000 to $3000.

6-3-1. County Surveyor may appoint deputies with consent of the County Court. Sec. 2 provides for removal of deputies.

6-5-1. The term of the County Surveyor shall begin January 1.

7-3-2. The Surveyor shall be furnished supplies and office space in the Court House. The County Clerk is custodian of survey records.

17-1-3. Defines a road, public road and highway. A road is established when it has been used by the public for 10 years or more, and public moneys or labor have been expended thereon, whether there be any record of its conveyance, dedication, or appropriation to public use or not. In the absence of any other mark or record, the center of the travelled way shall be the center of the road, and the right of way shall be an equal distance on each side, unless a different situation is recorded.

17-4-2. The State shall take over the county-district roads as of July 1, 1933.

17-4-7. The State Road Commissioner shall file with the County Clerk, all changes in titles to rights of way, maps, surveys, and discontinuance of state roads within the county.

18-9A-11. Provides for aerial photography of the State, for furnishing copies of tax maps and photos to counties and to the public. Opinion 51-78 (1964) of the Attorney General states that "if the State Tax Commissioner supplies a copy of the original tax map to a county, such map becomes county property and may be used in any manner the county officials concerned with property assessments may see fit".

30-13-1 et al. Relates to the profession of engineering.

30-13A et al. Relates to the practice of land surveying. See chapter on How to Become a Surveyor in West Virginia.

31-2-1 et al. Relates to surveys for railroads.

37-5-2. Surveyor for an adjoiner may go over neighbor's land to explore coal mines thereon to check mine location; not more than once a month.

37-12-1. When any surveyor is required to survey lands which have been surveyed before, he shall make such survey by the magnetic meridian, but shall also return and certify in his plat the degree of the variation of the magnetic needle from the true meridian, at the time of the resurvey; and also (if the same can be done) the degree of such variation at the time of the original survey; and every survey, whether original or not, shall be made by horizontal measurement, unless otherwise provided by deed or contract.

37-12-2. Provides for prompt survey and report if survey is ordered by the Court. Penalty of $50 to the injured party if survey is not made.

39-1-13, 14, 15, 16. Require subdivision plats to be recorded in the Office of the County Clerk and Assessor of the county wherein the land is located, before any lots are sold or offered for sale. Provides for a penalty of $200 for each period of 60 days that the plat is not filed, after notice is given by the county clerk. Plat of subdivisions in or abutting on incorporated municipalities shall be submitted to the council or commissioners of the municipality for approval, and such approval shall be indicated on the plat before recording.

39-1-17. Permanent corner monuments of stone or concrete, not less than 24″ long by 6″ square or in diameter at the top, with proper centers, shall be set at the most important corners of a subdivision of land; there shall be not less than two to each lot or block. Said monuments shall be shown on the recorded plat; no certificate of acknowledgement or other proof thereof is needed for recording such a plat.

51-3-3. The Circuit and County Court and Justices have concurrent jurisdiction over watercourses between their areas.

51-8-6. A free copy of each law, resolution, etc., of the Legislature shall be provided to the County Surveyor.

54-2-12. When land is condemned for use as a public road, a right of way only shall vest in the applicant, and not a fee title.

55-2-1, 1a. Adverse possession: must be actual, exclusive, hostile, open, and notorious, accompanied by a bona fide claim of title against all other persons, continuous for 10 years. Possession should be marked by fence, cultivation, residence, clearing, or any plainly visible manifestation. *Not enough* is a partial enclosure, timber cutting, or occasional grazing. Adverse possession of a private way over adjoining land for 10 years, known, acquiesced in, unobjected to, and unprotested by the owner, is presumed to be adverse to him, and ripens into an easement by prescription. A party may tack his possession onto possession of those under whom he claims. Possession carries against ancestors and heirs, even though infants. A squatter who makes no claim against the owner, develops no adverse possession. Ripened adverse possession is not destroyed by abandonment. Adverse possession against the surface owner does not affect severed minerals. An interlock exists when calls of one deed overlap calls of another. A call for the line of an adjoiner does not create an interlock; the only question is, where is the true line? If neither party has possession of an interlock, the older title prevails.

55-4. This deals with ejectment procedures. In this regard, Francis Silver V, an engineer and land surveyor of Martinsburg, West Virginia points out that the surveyor should know something of the law of "ejectment". Ejectment is an action at law to recover possession of real estate, perhaps with damages for wrongful withholding. Adverse possession may prevail only because the true or record owner has not seasonably asserted his right of ejectment. The surveyor should not abet a claimant by dignifying it unduly in his survey. The surveyor who strives to be just will find support in the law of ejectment. A plaintiff in ejectment proceedings must trace an unbroken chain of title to the grant or to a common source with the adjoiner who claims adversely. The surveyor, therefore, should consider the need to review the records back to a common source, and give proper weight to the senior line involved. Faulkner vs Thorn 120 W. Va. 575 and 122 W. Va. 323.

55-4-31. A proper petitioner may have the Circuit Court establish a true boundary line location.

59-1-6. Sets fees to be charged by the County Surveyor provided no agreement exists with clients.

59-1-10, 11. County Clerk's fees: recording plat with deed—$1.00, without deed—$2.00, unless plat contains more than 120 sq. in., in which event $0.007 for each additional sq. in. County Clerk's fee for providing copy of map—25¢.

61-3-30. It is a misdemeanor to break down, destroy, deface or remove any corner of land (tree or other monument). Penalty: $20 to $200 and 1 to 6 months in jail.

State Supreme Court Decisions

The following 3 cases cited and produced many key decisions of interest to surveyors:

West Virginia Pulp & Paper Company v. J. Natwick & Company, et al. No. 9216. 1941. 123 W. Va. 753.

Toppins v. Oshel & Copley. No. 10729. 1955. 141 W. Va. 152.

Blain v. Woods. No. 11089. 1960. 145 W. Va. 297.

A summary of each case is given in the W. Va. Law Review books, which can be found in most lawyer's offices. The significant "syllabus" points are shown at the beginning of each case. The main decisions cited or made in these and other cases are:

1. In locating land, the following are some of the rules resorted to, and generally in the order stated:

   a. Natural boundaries
   b. Artificial marks
   c. Adjacent boundaries
   d. Courses and distances (one controlling the other according to circumstances).

Neither rule, however, occupies an inflexible position, for when it is plain that there is a mistake, an inferior means of location may control a higher means. 38 W. Va. 1.

2. Quantity of land supposed to be conveyed in deeds must yield not only to natural landmarks, but to courses and distances. 32 W. Va. 487.

3. Where a deed contains a general description of land which can be made certain by proof of surrounding circumstances, or identified by reference to the land itself or other objects that more or less distinctly indicate or determine it, and such deed also contains courses and distances; such general description (if it satisfactorily appears from the deed itself or any recital or writing referred to therein that it was the intention of the grantor to convey the land thus generally described) will control, and the courses and distances insofar as they limit or differ from such general description, will be disregarded. 20 W. Va. 480.

4. Intent is the governing criterion, and a rule of construction that conflicts with the manifest intent of the parties is to be entirely ignored. 76 W. Va. 537.

5. Where the intent is clearly to convey a given acreage, quantity controls, and governs the location of a closing line between two named points to the extent of curving it instead of running it straight. 78 W. Va. 423.

6. A call of a deed to the line of an adjoiner in no case can make an interlock (lap) between the adjoining properties. A dispute between the two adjoiners does not of itself create an interlock. 123 W. Va. 753.

7. Calls in a deed for an adjoining tract are calls for a monument, and, where such tract is located for certain, it becomes a monument of the highest dignity. 81 W. Va. 76. (Also see 63 W. Va. 443 and 8 W. Va. 406.)

8. Monuments, natural and artificial, usually prevail over courses, distances, or mistaken descriptions in surveys or conveyances, or quantity of land. 123 W. Va. 753.

9. The location of lost ancient land monuments may be proved by reputation. 120 W. Va. 575.

10. If one end (A) of a disputed line in a deed is clearly fixed by evidence, and the other end (B) is unidentified and uncertain, and a line from (A) agrees perfectly with calls in a deed for course and distance, (A) governs (1) the location of (B), and (2) the distance called for in another line running to it. 85 W. Va. 415.

11. One or more monuments of a tract having been located, the courses and distances thereto are entitled to control others in said tract, the identifying evidence of which is slightly circumstantial and conflicting. 72 W. Va. 841.

12. A deed is construed as of its date, and a call in it for lands of another as a monument, is a call for the true location of such land at the date of the deed. 1011 W. Va. 711.

13. When coterminous owners, whose lines are disputed, obtained properties from a common source, and the lines of the junior grant run to the corners and lines of the senior grant, such lines of the senior grant are monumental lines. 142 W. Va. 404.

14. In an action under Chap. 55, Art. 4, Sec. 31, the party who for the statutory period has been in possession of the intervening acreage between the petitioner's line and the defendant's line may invoke adverse possession. 142 W. Va. 404. Petitioner has the burden of establishing his boundary by a preponderence of the evidence. 142 W. Va. 404.

15. Occasional and sporadic acts of dominion do not constitute adverse possession.

16. Color of title, as by adverse possession, is not to show a grant of land or interest therein, but to designate the boundary of the claim. 123 W. Va. 753.

17. A surveyor who is familiar with and has had experience with

monuments and courses of a survey may give his opinion as to identity of monuments. 72 W. Va. 449.

18. The opinion of a surveyor is not admissible as to true location of a boundary line. He may testify as to facts but not where he thinks a line is located. 69 W. Va. 346 and 145 W. Va. 297.

19. Testimony shall be discontinued if a surveyor testifies that a line was run at a certain place because the client says it was there. 128 W. Va. 383.

20. It is improper to prove an existing monument by reputation; only a lost monument may be so proved. 145 W. Va. 297.

21. It is inadmissible to allow witnesses to testify as to the location of a boundary line; they may do so only for lost monuments. 120 W. Va. 575.

22. It makes no difference whether a line is run in reverse or forward, so long as the calls conform to the deed. 145 W. Va. 301.

23. Where reservations are in a deed, the party who has the burden of going forward with the deed must also locate the reservation. 39 W. Va. 432.

24. A deed need not have courses and distances if the natural landmarks, artificial monuments, and adjacent boundaries can be located. 63 W. Va. 443.

25. The middle of Tug Fork is the boundary between Kentucky and West Virginia. 36 W. Va. 84.

26. Riparian proprietors of lands bounded on the Ohio River in this State, own the fee to the low water marks, subject to the easement of the public in that portion between the high and the low water mark. 29 W. Va. 314.

27. Once a highway, always a highway, unless abandoned or vacated in due course of law. 46 W. Va. 544.

28. A road is presumed to be 30 feet wide in the absence of proof to the contrary. 103 W. Va. 386.

29. It is not necessary that every part of a right of way be used and occupied. 124 W. Va. 233.

30. The edge of the right of way is determined by measuring 15 feet on either side from the center of the travelled way. 127 W. Va. 586.

31. There is a presumption that a "class A" road is to be 40 feet wide. 108 W. Va. 287.

32. A surveyor is apparently not entitled to file a mechanic's lien for his work. 117 W. Va. 160.

33. Locative calls control directive calls. 128 W. Va. 383.

34. Proof of oral agreement as to a boundary agreement was admitted. 131 W. Va. 218.

Miscellaneous Guide Lines

35. Lines run and marked on the ground are the true survey, and when they can be found, will control the calls for a natural or other fixed boundary and also are the boundaries in the grant where they differ from those produced by the courses and distances stated in the patent. 98 Penna. 37, 40.

36. It is generally held in most states that unless the State owns the fee, adjoining owners go to the center of the road, unless there are specific words to the contrary. A call *to* the highway, or *along* the highway, goes to the *center,* even if the distance reaches only to the edge. But if the call reaches only to the edge, and says "thence with the edge, line, or side" of the road, the boundary is the *edge.*

37. The above statements as to roads generally apply to non-navigable streams, small ponds and lakes (the state owning the bed of navigable streams). Slow changes in watercourses generally change the (riparian) boundary lines, but a sudden change does not. A call for the meanders of a stream usually goes to the center.

38. If a monument has width (such as a wall, stone, tree or ditch), the center is the boundary unless the deed states otherwise. 115 Me. 489.

39. A line dropped from the eave of a house is the line, when a house or corner of a house is called for. 113 Mass. 481.

40. Courses are presumed to have been run by magnetic meridian unless stated differently. 47 N. H. 235.

41. A fence may be the line or merely be one of convenience without intent to fix boundaries. 7 N. H. 436.

42. In the event of a line dispute or uncertain line, the adjoining landowners may agree where the lost line should be. The agreement should be reduced to writing and recorded. Consult an attorney and surveyor.

43. Instructions for U. S. surveys of public lands do not dictate how private surveys shall be made in this state. However, many Government methods are usable if state law does not conflict. Two methods from the 1947 MANUAL are mentioned here:

1. *The Dependent Resurvey.* When most corners of a survey are obliterated, then experience, thoroughness, and good judgment of the surveyor are indispensable for a solution to the problem. If the original survey was well done, and a good description made, the true location of corners can be determined within a reasonable limit of accuracy by a "dependent resurvey" (p. 319 MANUAL). Briefly this type of resurvey is one whereby existing evidence of the original survey is given primary control in locating corners and lines. This system is the one used for nearly all resurveys.

2. *The Independent Resurvey.* If, however, the original survey was grossly erroneous, or calls are fictitious, no rules can be laid down to solve the problem. When the original survey cannot be identified with any degree of certainty, and a survey by dependent conventional methods gives unsatisfactory results, an "independent resurvey" may be considered (Sec. 428 to 460, 1947 MANUAL). An independent resurvey disregards courses and distances in former deeds and may set up new calls which to the best of the surveyor's ability reflect the rights of all involved owners, especially as the intent is expressed or implied in the deed, or by possession on the ground. In case of dispute, the Court has final jurisdiction. The Chancery case of Hoffman v. Cain, No. 582 Randolph Circuit Court, was in part an independent survey. A resurvey may use dependent methods in one part, and independent methods in another part.

44. The following excerpts from an opinion on the JUDICIAL FUNCTIONS OF SURVEYORS by Justice Cooley of the Michigan Supreme Court will be helpful to all land surveyors:

> When a man has had a training in one of the exact sciences, where every problem within its purview is supposed to be susceptible of accurate solution, he is likely to be not a little impatient when he is told that, under some circumstances, he must recognize inaccuracies, and govern his actions by facts which lead him away from the results which theoretically he ought to reach. Observation warrants us in saying that this remark may frequently be made of surveyors.
>
> In the State of Michigan, all our lands are supposed to have been surveyed once or more, and permanent monuments fixed to determine the boundaries of those who should become proprietors. The United States, as original owner, caused them all to be surveyed once by sworn officers, and as the plan of subdivision was simple, and was uniform over a large extent of territory, there should have been, with due care, few or no mistakes; and long rows of monuments should have been perfect guides to the place of any one that chanced to be missing. The truth unfortunately is, that the lines were very carelessly run, the monuments inaccurately placed; and as the

recorded witnesses to these were many times wanting in permanency, it is often the case that when the monument was not correctly placed, it is impossible to determine by the record, by the aid of anything on the ground, where it was located. The incorrect record of course becomes worse than useless when the witnesses it refers to have disappeared.

It is, perhaps, generally supposed that our town plats were more accurately surveyed, as indeed they should have been; for in general there can have been no difficulty in making them sufficiently perfect for all practical purposes. Many of them, however, were laid out in the woods; some of them by proprietors themselves, without either chain or compass, and some by imperfectly trained surveyors, who, when land was cheap, did not appreciate the importance of having correct lines to determine boundaries when land should become dear. The fact probably is, that town surveys are quite as inaccurate as those made under authority of the general government.

Few, if any of the original stakes now remain. The stakes were often nothing but green sticks driven into the ground. Stones might be put around or over these if they were handy, but often they were not, and the witness trees must have been relied upon after the stake was gone. Too often the first settlers were careless in fixing their lines with accuracy while monuments remained, and an irregular brush-fence, or something equally untrustworthy, may have been relied upon to keep in mind where the blazed line once was. A fire running through this might sweep it away, and if nothing was substituted in its place, the adjoining proprietors might in a few years be found disputing over their lines, and perhaps rushing into litigation, as soon as they had occasion to cultivate the land along the boundary. If now the disputing parties call in a surveyor, it is not likely that any one summoned would doubt or question that his duty was to find, if possible, the place of the original stakes which determine the boundary line between the proprietors.

However erroneous may have been the original survey, the monuments that were set must nevertheless govern, for parties buy, or are supposed to buy, in reference to these monuments, and are entitled to what is within their lines, and no more, be it more or less. While the witness trees remain, there can generally be no difficulty in determining the locality of the stakes. When the witness trees are gone, so that there is no longer record evidence of the monuments, it is remarkable how many there are who mistake altogether the duty that now devolves upon the surveyor. It is by no means uncommon that we find men, whose theoretical education is thought to make them experts, who think that when the monuments are gone, the only thing to be done is to place new monuments where the old ones should have been, and would have been if placed correctly. This is a serious mistake. The problem is now the same as it was before: To ascertain by the best lights of which the case admits where the original lines were.

It will probably be admitted that no man loses title to his land or any part thereof, merely because the evidence becomes lost or uncertain. It may become more difficult for him to establish it as against an adverse claimant, but theoretically the right remains; and it remains as a potential fact so long as he can present better evidence than any other person. And it may often happen that notwithstanding the loss of all trace of a stake, there will still be evidence from which any surveyor will be able to determine with almost absolute certainty where the original boundary was.

There are two senses in which the word "extinct" may be used in this connection: one, the sense of physical disappearance; the other, the sense of loss of all reliable evidence. If the statute speaks of extinct corners in the former sense, it is plain that a serious mistake was made in supposing that surveyors could be clothed with authority to establish new corners by an arbitrary rule in such cases. As well might the statute declare that if a man loses his deed, he shall lose his land altogether. But if by extinct corner is meant one in respect to the actual location of which all reliable evidence is lost, then the following remarks are pertinent:

1. There would undoubtedly be a presumption in such a case that the corner was correctly fixed by the surveyor where the field notes indicated it to be.

2. But this is only a presumption, and may be overcome by any satisfactory evidence showing that in fact it was placed elsewhere.

3. No statute can confer upon a county surveyor the power to "establish" corners, and thereby bind the parties concerned. Nor is this a question merely of conflict between State and Federal law; it is a question of property right. The original surveys must govern, and the laws under which they were made must govern, because the land was bought in reference to them; and any legislation, whether State or Federal, that should have the effect to change these, would be inoperative, because of disturbing vested rights.

4. In any case of disputed lines, unless the parties settle the controversy by agreement, the determination of it is necessarily a judicial act, and it must proceed upon evidence, and give full opportunity for a hearing. No arbitrary rules of survey or of evidence can be laid down whereby it can be adjudged. The general duty of a surveyor in such a case is plain enough. He is not to

assume that a monument is lost, until after he has thoroughly sifted the evidence, and found himself unable to trace it. Even then he should hesitate long before doing anything to the disturbance of settled possessions. Occupation, especially if long continued, often affords very satisfactory evidence of the original boundary, when no other is attainable; and the surveyor should inquire when it originated, how and why the lines were then located as they were, and whether a claim of title has always accompanied the possession, and give all the facts due force as evidence. Unfortunately, it is known that surveyors sometimes, in supposed obedience to the State statute, disregard all evidence of occupation and claim of title, and plunge whole neighborhoods into quarrels and litigation by assuming to "establish" corners at points with which the previous occupation cannot harmonize. It is often the case that where one or more corners are found to be extinct, all parties concerned have acquiesced in lines which were traced by the guidance of some other corner or landmark, which may or may not have been trustworthy; but to bring these lines into discredit, when the people concerned do not question them, not only breeds trouble in the neighborhood, but it must often subject the surveyor himself to annoyance, and perhaps discredit, since in a legal controversy the law, as well as common sense, must declare that a supposed boundary line long acquisced in is better evidence of where the real line should be than any survey made after the original monuments have disappeared.

The mischiefs of overlooking the facts of possession must often appear in cities and villages. In towns the block and lot stakes soon disappear; there are no witness trees and no monuments to govern, except such as have been put in their places, or where their places were supposed to be. The streets are likely to be soon marked off by fences, and the lots in a block will be measured off from these without looking further.

Now it may perhaps be known in a particular case that a certain monument still remaining was the starting-point in the original survey of the town plat; or a surveyor settling in the town may take some central point as the point of departure in his surveys, and assuming the original plat to be accurate, he will then undertake to find all streets and all lots by course and distance according to the plat, measuring and estimating from his point of departure. This procedure might unsettle every line and every monument existing by acquiscence in the town; it would be very likely to change the lines of streets, and raise controversies everywhere. Yet this is what is sometimes done; the surveyor himself being the first person to raise the disturbing questions.

Suppose for example, a particular village street has been located by acquiscence and used for many years, and the proprietors in a certain block have laid off their lots in reference to this practical location. Two lot-owners quarrel, and one of them calls in a surveyor that he may be sure that his neighbor shall not get an inch of land from him. This surveyor undertakes to make his survey accurate, whether the original was or not, and the first result is, he notifies the lot-owners that there is an error in the street line and that all fences should be moved, say, one foot to the east. Perhaps he goes on to drive stakes through the block according to this conclusion. Of course if he is right in doing this, all lines in the village will be unsettled; but we will limit our attention to the single block. It is not likely that the lot-owners will generally allow the new survey to unsettle their possessions, but there is always a probability of finding someone disposed to do so. We shall then have a lawsuit; and with what result? Public policy requires that such lines be not lightly disturbed or disturbed at all after the lapse of any considerable time. The litigant, therefore, who in such case pins his faith on the surveyor, is likely to suffer for his reliance, and the surveyor himself, to be mortified by a result that seems to impeach his judgement.

Of course nothing in what has been said can require a surveyor to conceal his own judgment or to report the facts one way when he believes them to be another. He has no right to mislead, and he may rightfully express his opinion that an original monument was at one place, when at the same time he is satisfied that acquiscence has fixed the rights of parties as if it were at another. But he would do mischief if he were to attempt to "establish" monuments which he knew would tend to disturb settled rights; the farthest he has a right to go as an officer of the law is to express his opinion where the monument should be at the same time he imparts the information to those who employ him, and who might otherwise be misled, that the same authority that makes him an officer, and entrusts him to make surveys, also allows parties to settle their own boundary lines, and considers acquiscence in a particular line or monument for any considerable period as strong, if not conclusive, evidence of such settlement. The peace of the community absolutely requires this rule.

From the foregoing it will appear that the duty of the surveyor, where boundaries are in dispute, must be varied by the circumstances. (1) He is to search for original monuments, or for the places they were originally located, and allow these to control if he finds them, unless he has reason to believe that agreements of the parties express or implied have rendered them unimportant. (2) If the original monuments are no longer discoverable, the question of location becomes one of evidence merely.

It is merely idle for any State statute to direct a surveyor to locate or "establish" a corner, as the place of the original monument, according to some inflexible rule. The surveyor, on the other hand, must inquire into all the facts, giving due prominence to the acts of parties concerned, and always keeping in mind, first, that neither his opinion nor his survey can be conclusive upon the parties concerned; and, second, that courts and juries may be required to follow after the surveyor over the same ground, and that it is exceedingly desirable that he govern his action by the same lights and same rules that will govern theirs. On town plats if a surplus or deficiency appears in a block when the actual boundaries are compared with the original figures, and there is no evidence to fix the exact location of the stakes which marked the division into lots, the rule of common sense and the law is that the surplus or deficiency is to be apportioned between the lots on an assumption that the error extended alike to all parts of the block.

It is always possible when corners are extinct, that the surveyor may usefully act as a mediator between parties, and assist in preventing legal controversies by settling doubtful lines. Unless he is made for this purpose an arbitrator by legal submission, the parties, of course, even if they consent to follow his judgment, cannot, on the basis of mere consent, be compelled to do so; but if he brings about an agreement, and they carry it into effect by actually conforming their occupation to his lines, the action will conclude them. Of course it is desirable that all such agreements be reduced to writing; but this is not absolutely indispensible if they are carried into effect without.

I have thus indicated a few of the questions with which surveyors may now and then have occasion to deal, and to which they should bring good sense and sound judgment. Surveyors are not, and cannot be, judicial officers, but in a great many cases they act in a *quasi* judicial capacity, with the acquiescence of the parties concerned; and it is important for them to know by what rules they are to be guided in the discharge of their judicial functions.

45. Seniority. It is common practice in West Virginia to mention the names of adjoiners in a description of a survey or in a deed description, regardless of who has the senior and junior titles. The reason is for identification rather than for trying to assign seniority. This author believes that actual legal seniority must be based on other records or evidence rather than a statement in a survey description. The surveyor of course must determine seniority if conflicting calls exist. But in most cases it is the intent that properties adjoin on a common line rather than overlap or leave a gap. When a new parcel is cut out of a larger

tract, the new parcel is of course senior to the parent tract. Then calls for the adjoiner to the new parcel could state: "thence with land reserved by (grantor) xxx". But even if the description says: "thence with land of (grantor), the adjoiner is not senior; the facts show he is junior to the new parcel.

46. When riparian rights are involved, they appear to go to the low water mark. 26 W. Va. 644, 53 Am. Rep. 116; 35 W. Va. 95, 13 S.E. 42, 29 Am. St. 793; 54 W. Va. 122, 46 S.E. 269, 102 Am. St. 927; Md. vs W. Va. 217 U. S. 577 (1910), 54 L. Ed. 888, 30 Sup. Ct. 630.

47. The western line of Maryland which joins West Virginia is to be run and located and marked "as far as practicable as it has been generally recognized and adopted by the people residing about or near the same xxx and not as conforming, except to a limited extent, to the western boundary of the Maryland Military Lots as said lots are now located and held", Maryland vs West Virginia 217 US 577 (1910).

Fences as Boundary Lines

One of the most common kinds of evidence that may show property ownership is a fence. In urban areas fences are made of wood, stone, concrete, metal or shrubbery. In rural areas fences are usually wire (barbed or woven) or rail. A ditch or ridge of earth (as from plowing along a fence) may have significance. The early zig-zag rail fences are being replaced by wire. In West Virginia, the law requires adjoining rural owners to cooperate in fence construction and maintenance where the fence serves certain agricultural uses of each owner. Fences assume considerable significance, depending on circumstances.

The good surveyor will tie his random line to nearby fences at about every station set-up. So when the random line is plotted, he can plot the fence and know how it fits between corners. Measurements are made to the wire or center of rail fences. The fence can sometimes serve as a guide to where the random line was, in the event random stakes are rotted off, lost or removed. When a fence is the legal boundary line, the wire itself rather than the center of the posts, is the proper line. Some courts have held that where the center of the posts have been placed on line, the fastening of wires on one side of the posts deprives the owner on that side, of use of a part of his land.

If fences are found to coincide with measurements in deeds for the properties concerned, they create no problem for the surveyor. However, when fences are found that diverge, the significance of the fence must be determined. A monument called for and found at a corner apparently undisturbed will usually take priority over a fence line or intersection. But where no monument is found, an old fence intersection in the area raises a presumption that the corner was at or near the intersection, especially if there is no evidence from adjoining owners to the contrary. The older the fences, the greater the weight that is accorded them by the courts.

Fence intersections frequently provide a good guide to corner locations because more care is exercised by the builders and markers were often present. But this is not so true for fence lines *between* corners. When a fence is straight between recognized corners, no problem is raised by its presence. Many fences, however, diverge a great deal from a straight line, but if the corners at each end are agreed to be correct, the owners are likely to agree that the fence location is one of convenience only, and was placed there because it was cheaper to build off line, or because the line was unmarked and its location was not definite. If a resurvey is to be made, the owners will likely agree that the fence does not control and will be relocated to the true line where possible, at some future date.

A fence on the bank of a stream or road is usually there for convenience; the true line may be the center of the stream or road. Where a fence is called for in a deed, it controls if in its original location.

Disputes are most likely to arise when markers are missing, when measurements are inconclusive, and fences not straight. The surveyor must search carefully for evidence of fences that may have existed before the present fence. Chestnut rails are long-lasting and often their decayed remains can be found. A boundary line agreement may be found, either implied or formalized, in writing or verbal. If the fence has been acquiesced in for over 10 years (in West Virginia) and the land on each side has been used continuously by the respective owners or their tenants, a strong case exists for considering the fence to be the correct line. When such fences are "ancient" (thirty to fifty years old) and there is no good evidence to the contrary, courts are likely to hold that the fence is the line by possession or prescription. But in this connection, surveyors need to know about the legal action of "ejectment".

The original survey and grant presumed a straight line between two corners to be the true line. However, if the original surveyor's "footsteps" can be found elsewhere (by marked trees for example), the "footsteps" will control. When no such footsteps can be found (and very often they cannot), the modern surveyor should attempt to obtain agreement between the owners to accept a straight line between the corners as most equitable to their respective interests. If no agreement can be obtained, an off-line fence must be tied to and a lap shown on the plat; the ownership of the lap is a legal problem. There will be few times when a fence controls the line location, and surveyors must be alert to recognize them.

The surveyor may occasionally be asked where a fence should be built. Some owners prefer to build fences entirely on their own land in the hope that it will save an argument with the neighbor. However, it is usually better in the long run if fences are built *on line* or as close thereto as practicable. If a fence is deliberately set back from the true line, and later one or both monuments are lost, the fence can cause disputes and require extra survey costs to unravel the problem. And the owner who had set the fence back might be hard pressed to prove his ownership to the true line.

Old Line Marks as Boundary Controls

Most land grants in West Virginia state "according to survey made on ———— by ————" and by authority of Land Warrant #———— for ———— acres. Therefore, it must be presumed that most grants were preceded by a ground survey. But when a grant was to be issued for land adjoining property previously surveyed and granted, it was likely customary to survey only those lines of the new grant not previously surveyed, and to incorporate in the new grant the calls for the line of the previously surveyed grant or survey. This practice gave rise to significant errors of closure.

The lines of senior grants take priority over later grants. As most original grants in West Virginia were made 100 years ago or more, and the lands involved have been cutover one or more times or cleared of all trees since then, it is apparent that most trees if marked by the original surveyor have long since disappeared.

Original grants have been divided and subdivided by deed or otherwise right down to the present time. So any survey that subdivides land not heretofore subdivided is an "original" survey. (A resurvey of a former survey is not an original survey.)

The courts hold that original marks of a survey are monuments of the highest dignity because they show the "footsteps" of the original surveyor. Therefore, original marks control location of a line if:

1. The survey is senior to the adjoining survey.
2. The survey was made before or simultaneously with the execution of the deed or grant.
3. There is no expressed intention to the contrary.
4. The line is not a meander line (as on the bank of a stream or road).
5. The line is not a random line (which should never be marked).

A resurvey must follow the original survey if its footsteps can be found even though the original survey line was crooked or apparently on the wrong bearing and distance. Original marks can be hacks or blazes on trees, stakes set at instrument setups on the true line, paint marks, and topographical features such as stream and ridge crossings. The marks must be identifiable. The nearer they are to a corner, the greater their weight. The best way to prove the identity of a mark on a tree is to block it out, or remove an increment core, and see if its age checks with the date of survey. Other ways include the testimony of witnesses who saw the mark made. The age of stakes or paint is more difficult to tell.

It is understandable how time has erased nearly all grant survey marks. But it is logical that a surveyor now should find marks on trees of original surveys of *recent* times. However, the author has not found many recently marked lines of surveyors in the woods. It appears that many original surveys have been made without trees being marked, perhaps because the client did not wish to pay for the added cost or because it was believed the brush cut on the random or true line would suffice. At any rate, the modern surveyor who is retracing original footsteps unfortunately may not find many marked lines.

Marks found along a random line must be tied in and plotted to see if they approximate a line that would have been made by a surveyor. Several marks should be blocked out or bored to check their age. Factors used in considering weight to be given are: whether a surveyed line was called for, the age of the marks, their frequency and relation to known corners, and whether they are center or sideline marks. Center line marks are relatively rare, and this fact allows the surveyor some leeway when marks are chiefly on sideline trees.

When a survey line crosses topographic features at or near a right angle, the feature does not provide as good an indicator of line location as does a line crossing at a smaller angle. In the former case, a passing distance could fit the feature at more than one place, but in the latter, the distance narrows the location greatly.

When corner locations at the ends of lines are vague, a straight line that fits a majority of the line marks will indicate that corners should be reestablished at or near the intersection of the marked line and the adjoining lines. When authentic corner markers exist, but the marked line does not intersect with them, the best compromise should be selected, keeping in mind the dignity assigned to original marks. Judgement is the guide; it is difficult to be more specific.

From the previous discussion, it should be apparent that every surveyor who is to make an "original" survey, should be careful to use that precision and choice of methods that will meet the needs of the next generation and the rules of the present. And, of course, the new line should be well marked. When a resurvey finds no "original" marks, the resurvey marks are likely to be considered in the future as an original survey, although legally they are not.

Chapter 3

SURVEYING RURAL LAND ALONG TRAVELLED WAYS

Chapter 2 lists certain laws and court decisions relating to public roads in West Virginia. This chapter describes how surveys or resurveys of private property boundary lines should, with few exceptions, be conducted when these lines adjoin or cross a travelled way in this State. It will not cover ways over large bodies of water; they are not common here. However, the right of access to a stream or spring on another's land may need to be located by the surveyor. Travelled ways are private or public; sometimes it is difficult to tell which type is involved.

Private Ways

Private ways consist of paths or lanes usually leading from public roads. Sometimes an old deed will refer to a path, which might be helpful if identifiable. A "way" may be a right of another to go over a client's land, or a right of a client to go over another's land. "Ways of Necessity" are discussed in a booklet entitled *Rights of Way for Removal of Natural Resources from West Virginia Land* published by the West Virginia Center for Appalachian Studies and Development, August 1, 1965, as Public Affairs Series No. 1, West Virginia University, Morgantown (free).

Crossings. When the line of a survey crosses a way of some significance, the surveyor should record the distance to the estimated center, perhaps its bearing, and any other relevant data. He will decide later whether to mention it in the description of survey. That decision might hinge on whether the way has legal status or whether it is needed as a monument along the line. If it might be a public way, it should be included. The way may first be noted on the random survey. But the crossing point desired is where the true line will cross it; this point may be found in several ways:

1. If the path is straight, its bearing may be taken at the random line crossing and its position on the true line found by platting.
2. After the station on each side of the way is offset to the true line, the distance to the way center may be measured.
3. If the random and true line are close together and cross the way at a right angle, the random horizontal distance can be considered equal to the true line distance.

The description can read: "xxx N 12-43 E, at 12.85 poles crossing a lane bearing about N 20 W, in all 24.56 poles to xxx."

Corners. If a way forms the boundary of a survey, a monument must be placed (if not found) where the true line enters and leaves the center of the way, or a witness corner established. A flat-topped stone with letter on top set flush with the ground in the center of the way and properly referenced would be good. If the marker could be disturbed by traffic, a reference marker should be set in the true line on the edge of the way where it is not likely to be disturbed. The true corner or witness marker should be well referenced.

Surveys Along Private Ways. A survey can be made in two locations:

1. Generally along the middle of the way. The front rodman selects the next station where he can see the compassman *and* the route ahead, without trying to get any specific distance other than as long as is reasonable, and without changing grade appreciably. A station may be several feet from the middle of the way to avoid short shots; then the notes might show the distance to the middle and to the point where the line of sight leaves and returns to the middle. Nails and bottle caps make good markers for key stations, an X in the dirt for others. The compass may be set on every other station and a backsight taken; this saves time but eliminates the chance to check for errors or magnetic attraction. This "leap-frogging" is not recommended except perhaps for very rough reconnaissance.

2. The former courses can be resurveyed after allowance for variation. If the bearing selected for the first shot appears not to hit the proper line, adjust it and try again. If satisfactory, use the same adjustment on the other courses. If the survey does not end at the point called for, the bearings and distances can be adjusted as in the Chapter on True Lines. If the difference is small, the last course should be adjusted. Remember that although the deed said "with the road", the surveyor may have "cut corners" and not followed the exact middle. If the surveyor makes new calls, and the adjoiner also calls for the way as a line, there can be no legal disagreement, because the road is higher in priority than courses and distances. The new description must say "with the lane" or "with the center of the road". If private ways change location due to detours, calls can help find which route is the true boundary.

Public Roads

Historical Background of West Virginia Public Roads. In 1785 the General Assembly of Virginia first authorized roads to be built in this area. The first was from Winchester, Virginia to Morgantown. From 1863 to 1873 County Courts authorized construction. In 1909 the Legislature established width standards. In 1910 the only paved roads were 2 to 4 mile sections out of Wheeling, Huntington, Clarksburg, and Parkersburg, and a few miles of toll roads. In 1911 the first concrete road 1½ miles long was built out of Williamstown in Wood County.

Until 1920 the intercounty or main roads were called "Class A" roads and received state aid; other roads were "Class B". Two Class A routes were allowed each county. There were 1200 miles of paved roads but only two county seats were connected by them. In 1921 the State began to build roads and by 1933 there were 7423 miles of surfaced roads; between these years there were county *and* state roads. In 1933 the State "took over" the County District roads and called them "Secondary" roads. Some public roads of lesser importance remained under control of counties and towns.

The 1967 Legislature reclassified the roads according to the service they provided: Expressway (major intra- and interstate travel), Truckline (major city to city), Feeder (between communities) and Local Service (access to abutting land).

Before surveying along or across public roads the surveyor should determine three things:

1. What type of title exists in the public
2. What is the width of the right of way
3. Where is the center (or edge) of the road.

Title. Public right to use roads stem from these sources:

1. Ungated, uncontested and continuous use (in West Virginia, for more than 10 years). If public money or labor has been expended for maintenance, they are by law (17-1-3) presumed to be fully public. There is seldom a record title for this type of road, but litigation can be found in the Circuit Clerk's Office.

2. "Minute Orders" may be found in the County Court records before 1933, establishing a route as public, ordering a road to be maintained, awarding damages to private parties, or appointing commissioners to go upon the land and assess damages. Search the "Order Books" under name of property owners affected.

3. Deeds in Deed Books in the County Clerk's Office. The Grantee may be the County Court or the State Road Commission. Since 1933 it has been the practice to obtain title for new roads by deed.

4. Condemnation orders in the Circuit Clerk's records indexed under plaintiff and defendants, the latter being more helpful to the surveyor. Orders usually result in recorded deeds.

Width of Right of Way.

1. For roads established before 1933 no written record of width will likely be found, unless it be for main roads like US numbered routes. The rule is that (in the absence of documentary evidence to the contrary) Class A and State roads are 40 feet wide (20 on either side of center of the travelled way); Class B and County District roads are 30 feet wide (15 on either side of center). Court cases supporting these widths are: *State Road Commission* vs. *Coogle,* 108 W. Va. 287;

Childers vs. *State Road Commission,* 125 W. Va. 233; *Hark* vs. *Lumber Company,* 127 W. Va. 586.

2. For roads established since 1933, a document will usually be found. In recent years it has been the policy to acquire easements or fee title for widths of 40 feet or more. When no document exists, roads are presumed to be public under Code 17-1-3. However, no presumption of a specific width exists, and the right of way is limited to the land actually occupied by the road, including shoulders, cuts, slopes, fills and drainage. *Dunn* vs. *Griffith,* W. Va. 82 SE 2nd 300.

Where is Road Center or Edge. The center of a road is ordinarily the center of the travelled way. It is usually midway between the edges of the blacktop or concrete surface. Edges are sometimes irregular; take the average of several measurements at the more uniform places. Sometimes the surfacing is widened on only one side of center and this must be allowed for. Where curbs exist, measure between the square outer edges. Check distance between fences. Use an average of as many methods as possible.

If a document or plat shows that the right of way is not of equal width on both sides of center, the document would control; this is likely to be the case for main roads in recent years. Plats often show reference markers (H/T or nail and tack in trees) but these may not last long enough to be used in later surveys. Unfortunately, permanent markers are scarce along roads.

If research fails to reveal where the edge or center of a road is located, the surveyor may decide to ignore the issue. This can be done by establishing a marker as best he can and describing it as being "on or near the edge of the right of way" or "at ———— feet from the approximate center of the road."

State Specifications for Markers. The 1960 Standard Specifications for Roads and Bridges in West Virginia in SECTION 2.129 lists the policy and methods of installing markers as follows:

> "2.129.1 *Description:* These items shall consist of furnishing and placing at the points designated on the Plans or by the Engineer, of precast concrete Project Markers, Right-of-Way Markers, and/or Survey Markers, in the manner and subject to the conditions and regulations hereinafter prescribed.
>
> "2.129.2 *Materials:* xxx The right-of-way markers and survey markers shall have a metal plate on top of the design and conforming to the Specification shown on the Plans.
>
> "2.129.3 *Construction Methods:*
>
> (A) Manufacture of Markers: The markers shall be cast to the dimensions shown on the Plans.
>
> (B) Installation of Markers: Project markers shall be set at the

beginnings and endings of projects, at points designated by the Engineer. Right-of-way markers shall be set at points designating the right-of-way boundaries. Survey markers shall be set as shown on Plans or as designated by the Engineer. All markers shall be set plumb and to the depth shown on the Plans. The holes shall be backfilled with suitable material thoroughly tamped. Right-of-way markers shall be accurately set to the mark given by means of intersection strings. If approved by the Engineer, other equally accurate methods may be substituted."

State Road Plats and Descriptions. The State Highway Commissioner is required by law to file with the County Clerk a copy of the plat for all road acquisitions, changes, and abandonments. Plans usually contain (a) plan by horizontal survey, (b) profile or vertical elevation, and (c) cross-sections. Plats are also on file in the District Engineer's Office but these may differ, in which case the Clerk's copy controls.

The deed description for road parcels usually shows widths, stations in feet along centerline, and curves according to radius, or length left or right of center in the direction of stationing. Some older descriptions merely called for conveying a right of way "as shown on the plat." Here is a sample description:

> "Beginning at a point in the north r/w for Project ——— 30 feet left of center at Station 146 + 22, thence with a curve to the left of a radius of 1232 feet, about 216 feet to a point 30 feet left of Station 148 + 38, thence on a radian line from said center line 20 feet to a point 50 feet left of said centerline, thence with a curve to the right with radius of 1212 feet about 500 feet to a point 50 feet left of Station 153 + 38, thence S 32-04 E 896 feet parallel to centerline to a point 50 feet left of Station 162 + 34, a point in the property line of ——— and ———, thence with a curve to the left xxx."

Some descriptions call for "additional widths for cuts and fills"; some include temporary or permanent easements for road-related purposes. Survey textbooks show symbols used on road plats for simple curves. The significant ones for plotting are: central angle, radius of curve, and length of chord or arc.

The present trend in road descriptions is away from center line controls, and toward straight bearing and distance boundaries that are more easily located and moumented, and mean more to the adjoining property owners and private surveyors.

Federal and Interstate Highways. The State usually acquires title for these highways. The 1956 Federal Highway Act authorizes payment to States for establishment of markers.

National Forest Roads. There are three National Forests in this

State—the George Washington, Monongahela, and Jefferson. The Forest Service builds roads on Government land, and private land near the Forests. They are generally open for public use, but sometimes are subject to certain restrictions, and may be closed for forest fire or other emergency. The U. S. Government usually obtains easements over private land rather than fee title. They call for ———— feet on each side of centerline as surveyed, or as shown by plat on file in the Supervisor's Office, with added width for cuts and fills.

Surveys Across or Along Public Roads

The surveyor normally does these preliminary jobs:

1. Abstract the records in the Court House.
2. Examine the County Road maps.
3. Obtain data from Road offices.
4. Make tracing of road plans covering the area of interest.
5. Obtain information from client or neighbors.
6. Inspect the road on the ground.

Crossings. Property lines cross roads only when the road is an easement and not a fee title. The surveyor should locate the center or edge, and establish a witness marker on the edge or at some safe place, in the line. Tie to the road center and any nearby visible corners. The same methods described under Private Way Crossings can be used. If there is a State Road Survey, the surveyor should try to tie his survey to some point on the State plan or easement. If this can be done where the road enters and leaves the property, the road can be shown on the client's plat; include the width, its Project No. and station references.

If a boundary crosses a road at other than a right angle, the edges are found by measuring from the road center at right angles to where the boundary enters and leaves. The length of the boundary crossing at a road tangent may be checked by plotting, or by use of a right triangle formula. Crossings at curves may be checked by plotting on a large scale.

Corners. When a road forms the boundary being surveyed, there will be at least two corners at the road—where the line enters and leaves.

1. If the road is an *easement,* the boundary will usually run to the center. A N/C or spike can be set at the center, and a witness marker set on the edge or where it will be as permanent as possible. Reference the center or the witness marker, including a tie to the plan. The survey should be made along the road center if traffic permits. If traffic is too heavy, the survey should be made on the shoulder or edge. Take backsights to check for magnetic attraction which is common along roads. Run the distances and (adjusted) courses of the previous survey, or use new meander courses. The final shot must tie to the center of the road

and bearings adjusted as under *Private Ways.* If there are no calls, the surveyor should make new ones.

The description of such a survey could be: "xxx S 60-10 E——— to a point in center of a public road, from which a 10" red oak witness is N 60-10 W 31.1 feet; from said witness tree, a ——— bears ——— ——— feet thence along the center of said road S 10-25 W 14.2 poles to a point in center of the road, corner to ———, from which a set stone marked with an ——— bears N 44-50 W ——— feet, thence leaving said road, N 44-50 W xxx." The legal boundary of the land is the road center and not the courses.

2. *Fee* roads. If the property lies on one side of the road, a marker should be established (if none is found) on the legal edge. The survey along the road can be made as described above, but adjusted to fit the edge of the road as described in the road deed. If the road cuts the client's land into two or more parcels, the entire property is usually surveyed as a unit, then corners established (if none found) on the legal edge where the lines touch on both sides of the road. The road is traversed as before described. Each parcel is described separately. Calls along the road must be compatible with the road description.

Abandonment of Roads

Abandonment of private roads seldom creates any survey problem. Abandonment of public roads is primarily a legal problem. Many years ago, abandonment was not a very formal matter in West Virginia: use ceased, maintenance stopped, perhaps gates were erected and not disputed. The law now provides a formal procedure for abandoning public roads and streets. When a road is abandoned, the surface may revert to the current owners of the abutting land, but here again this is a problem for the attorney.

Chapter 4

ETHICS FOR LAND SURVEYORS

The practice of land surveying in West Virginia has far to go to reach a professional level. Lawyers, doctors, and engineers have attained professional stature by self analysis and hard work. To proclaim oneself as a professional surveyor does not make it so; it is so only if society accepts surveyors as a group as professional. Only as each surveyor raises his own quality of work, will the group be benefitted. Surveyors should have a code of ethics to guide their actions. They cannot do better than to follow the Code of Ethics of the National Society of Professional Engineers. An engineer member pledges to:

1. Dedicate his professional knowledge and skill to the advancement and betterment of human welfare.
2. Give the utmost of performance.
3. Participate in none but honest enterprise.
4. Live and work according to the laws of man and the highest standards of professional conduct.
5. Place service before profit; the honor and standing of the profession before personal advantage; and the public welfare above all other considerations.

The following guide lines should help the surveyor; they are adapted from the Professional Engineers' Code of Ethics:

1. Admit an error if proven obviously wrong.
2. Refuse a job if it is not in accord with law, ethics, and sound survey principles.
3. Avoid conflicts of interest or business connections between clients; inform interested parties if a conflict appears to exist.
4 Use reasonable safety precautions for himself and his men.
5. Do not sign for another's work unless the signer accepts personal responsibility for the work.
6. Do not advertise work in a way that might bring discredit to surveyors in general or another surveyor in particular. A surveyor's own work is his best advertisement.
7. Do not tell confidential information of a client without his consent.
8. See that all non-confidential matters of his survey are made available for public use.
9. Undertake only those jobs for which he is qualified by education, training, and experience, referring to others who he believes qualified, those jobs which he cannot handle.

10. Uphold the principle of adequate compensation for surveyors. Do not perform free services unless perhaps for civic, charitable, religious or nonprofit organizations. Do not charge fees substantially below accepted standards of other surveyors in the area, considering the amount and quality of the work to be done. Do not charge less for less difficult work.
11. Do not accept compensation from more than one client for the same work unless all parties know of the situation.
12. Bring errors to the attention of the surveyor responsible, and if not corrected, to the attention of the clients involved. Correct his own errors resulting from negligence or carelessness, usually at no cost to the client. Do not associate professionally with surveyors who do not conform to ethical practices.
13. Do not bid competitively for a job except to state his regular fee schedule.
14 Do not review the work of another surveyor for the same client, except with the knowledge or consent of such surveyor, or unless the client has severed his connection with such other surveyor.
15. Give credit to other sources of information and data. Abide by patent and copyright laws.
16. A client has a right to examine and reproduce a surveyor's data pertaining to the client's project.
17. Cooperate with other surveyors by exchanging non-confidential information and experience.
18. Discuss conditions of hire with helpers in advance of work.
19. Exercise independent judgment, yet be able to support work with facts. Evidence from people must be weighed in the light of evidence found on the ground or in the records.

The policy on advertising by means of cards, yellow page and newspaper listings, vehicles, signs and gadgets, as approved by many surveyor's and engineer's associations of the country, are listed for the guidance of those who may have a question concerning the propriety of their advertising.

Business cards and letterheads may denote the firm name, address, telephone number and the type of work the firm engages in; the individual's name and title; and the seal or name of the surveyors organizations of which the individual is a member. It is considered undignified to state years of experience, errors or omissions insurance or the license numbers for Land Surveyor.

Yellow page listings should consist of a single line, stating the firm name and address, and telephone number. Display advertising is considered undignified.

Magazine and newspaper advertising is considered improper mediums of advertising for services. Business card listings in charitable, educational or professional journals is considered dignified.

Project directory type jobsite signs listing the firm name and city, and containing the type of work being done is considered dignified and professional.

Identification on surveying vehicles should be limited to the panel of the door, and should state only the firm name and city, the words "Land Surveyors" and the telephone number.

Billboards, radio and television advertising is considered unprofessional and undignified.

Office signs should be in good taste and of an appropriate size listing only the firm name, and type of services offered. Licensed personnel may also be listed.

Calendars, gifts and gadgets such as cigarette lighters, key chains, pencils and pens are considered undignified professional advertising.

Leaflets and circulars are regarded as an undignified method of obtaining surveying work. Association pamphlets and educational type circulars are commendable, however, they should not be used by individual firms for advertising purposes.

A Profession

"A profession has six fundamental characteristics. First, a unique body of knowledge and advance training is required; emphasis is on mental rather than manual tasks. Second, a profession deals with service towards mankind involving considerations beyond earning a livelihood. Third, the accretion of knowledge through research must be passed on to the practitioner. Fourth, members of the profession form an association which usually enforces standards for the profession, advances and disseminates information, and screens entry into the profession. Fifth, a test is often required to indicate the competence of the individual, sometimes accompanied by licensing. Sixth, there must be ethical guides for how the practitioners should act among themselves or in relation to society."—by J. W. Barrett, Department of Forestry, University of Tennessee

Professional stature is the respect of society. It is a way of life. It is bestowed gradually, as it is earned by an individual. When enough surveyors, individually, gain the respect of all with whom they have dealings, both in their practice and in their private lives, surveying will be a true profession. The practice of the "Golden Rule" in every action, public and private, will gain that respect that is professionalism.—by Chuck Wooldridge, President California Land Surveyors Association

"There is only one criterion for a professional, and that is competence. The total society requires this specialty. If competence breaks down, the profession breaks down."—by Dr. George L. Farre, Georgetown University

Chapter 5

LAND SURVEYING AS A CAREER

It is important that surveyors be qualified by character, ability, and training. This chapter summarizes what it takes to be a good land surveyor.

Education and Training

The potential surveyor should be a high school graduate, be average or better in mathematics, and must have taken algebra, geometry and perhaps trigonometry. From here on there are several routes: one is to graduate from a two-year college course in surveying technology. Another is to graduate from a four-year college course in surveying. A third way is to take one of the correspondence courses now given in the United States. A period of training and work under a practicing surveyor is essential; the length of the period varies with the ability and education of the trainee.

Characteristics

1. A surveyor must above all be dependable, have good character and sobriety; his morals must be above reproach.
2. He must have the ability to reason, to analyze, and to draw conclusions from the evidence and the law.
3. He should be average or better in English because there will be many reports to be written.
4. He should be in fairly good health. Eyesight should be normal or better; glasses are seldom a hindrance. He should have a sense of direction and proportion. He should be willing to work long hours when needed.
5. He should have a curiosity to know—and the ability to investigate—all facts bearing on a situation.
6. Lastly, he should be able to deal tactfully with people.

Financial Income Expected

The money to be earned from land surveying will depend on the ability and energy of the surveyor. Most surveyors and their helpers receive adequate income. Rural surveyors earn less than city surveyors but their expenses are less.

Sources of Employment

Many companies—coal, oil, gas, lumber, electric—have surveyors on their payroll. Federal, state, and city government agencies employ surveyors. There is a great need for one or more full-time surveyors, self employed, in each rural-type county. As owners recognize the need for adequate surveys of their property, the demand for land surveyors will rise.

How to Become a Surveyor in West Virginia

The laws of most states describe the requirements for a land surveyor to practice. In 1969 the Legislature of West Virginia enacted into law Chapter 30, Article 13 A; it covers cadastral, underground, and hydrographic surveys. To be eligible for a license to engage in the practice of land surveying, the applicant must:

1. Be at least 18 years of age
2. Be of good moral character
3. Have been a resident of the United States for one year immediately preceding the date of application
4. Not have been convicted of a crime involving moral turpitude
5. Have 4 or more years experience in land surveying under supervision of a licensee of any state (or a person eligible for such license), provided that up to two years of satisfactory study in an accredited surveying curriculum may be substituted for experience, and
6. *Pass an examination* on land survey knowledge, skills, and techniques.

Eligible for a license *without examination* are:

7. Licensed surveyors from other states whose requirements are as great as those of West Virginia.
8. Graduates of accredited surveying curricula plus 2 years experience under a licensee of any state (or eligible for such a license), provided items 1, 2, 3, and 4 above are met.

Exempt from licensing are: professional engineers registered under W. Va. Code, Chapter 30, Article 13; employees of persons, firms, associations, corporations, U. S. Government, the State of West Virginia, or any political subdivision thereof, when such employee is engaged in the practice of land surveying *exclusively* for said entity.

Four Kinds of Surveyors

1. The Land Surveyor who doesn't know and doesn't know he doesn't know. This man will not learn and cannot be helped—ignore him for he is a clod.
2. The Land Surveyor who knows—but doesn't know he knows. This man can learn—help him for he lacks confidence.
3. The Land Surveyor who doesn't know and knows he doesn't know. This man is eager and willing to learn—he must be helped for he is a jewel.
4. The Land Surveyor who knows and knows he knows. This man has learned and needs no help—get out of his way.—from Wisconsin Society of Land Surveyors, March 1970 Newsletter

Chapter 6

BOUNDARIES FROM A LANDOWNER'S VIEWPOINT

Where is the Line?

Each landowner should know where his property lines are located. He should be able to stand anywhere on the lines and be within a foot or two of the correct location. There should be no disagreement with his adjoiners. The corners and lines should be clearly marked so that no encroachment by a reasonable person is possible. Each corner should be a tree, pipe, stone, or other substantial marker, adequately referenced by witness trees or other semi-permanent objects.

What the Owner Can Do

The reestablishment and marking of corners, witnesses, and lines should be done *by a surveyor* or under his direct supervision. The landowner, however, may improve his boundaries in some ways: by repainting over old paint at 8-10 year intervals; by straightening or refirming loose monuments; by placing concrete around iron monuments. Disputes with adjoiners should be worked out peacefully and tactfully, recognizing that compromise may be necessary. An owner should not prepare a new description of his land for a deed.

When is a Survey Needed?

When the conditions in the first paragraph above are not met, the owner should discuss the situation with a reputable surveyor. He will likely make no charge for a preliminary talk. A resurvey may not be needed. A good time to analyze the problem is when:

1. Timber is to be cut on owner's or adjoining land.
2. Land near the line is to be cleared.
3. The property of the owner or adjoiner is to be sold.
4. An estate is being settled.
5. When only 2 or 3 corners of a property are visible.

Selecting a Surveyor

Select a licensed land surveyor with a good reputation. Do not assume that all engineers are land surveyors. An engineer may be capable of surveying a city lot but not lost rural property lines. Neither is a land surveyor an engineer. North Carolina Rules for Land Surveying state: "It is not anticipated that a Registered Engineer will endeavor to practice any aspect of land surveying unless he is adequately qualified through experience and training in addition to his formal education."

Surveyors differ in their ability or preference to perform various kinds of their work, just as do lawyers, engineers, or other professional people. Here are some different kinds of surveys:

1. Rural, farm or mountain boundary lines.
2. City residential lots.
3. Subdivisions.
4. Highways, rights-of-way, utility lines, etc.
5. Mining—underground or surface.
6. Grade lines for buildings, sewers, water lines, etc.
7. Commercial or industrial property.

Some surveyors are good at diagnosing boundary disputes and solving them, while others do not want to "get mixed up" with lawsuits. Some work with a transit, some with compass. Some have made surveys near the land in question and know where adjoining corners or key monuments are located. Some may have to travel long distances to get to the job and therefore must charge more. The person who has a survey problem should obtain a surveyor well qualified for the job at hand—one who:

1. Is qualified by training and experience.
2. Has the proper equipment.
3. Can do the job within the limit of time agreed upon.
4. Is licensed or registered by a State regulatory agency.
5. May have made other surveys in the area.
6. Will testify impartially in court if there is a dispute.

How Can a Good Surveyor Be Found?

1. Search the yellow pages of the telephone book.

2. Ask others who have had experience with surveyors. These include attorneys, county officials, or friends. Employees of Highway Departments, utility and other companies often survey privately part-time.

3. The deed or plat books in the recorder's office is a good source. Start at the latest recordings and work backward. Watch for name, address, and seal on plats. A good surveyor will show on his plats the data described in the chapter on plats; the description in accompanying deeds will be comparable to the example in the chapter on descriptions.

4. If there is a dispute or other unusual aspect to your job, find out from others who have had similar problems, whether the surveyor furnished them with a Certificate telling them how their survey was made.

5. Inquire of the State Board that licenses land surveyors.

6. Inquire of the local or state Association of Land Surveyors.

Do not ask surveyors to bid competitively, nor select one just because he quotes the lowest cost or rate. If one surveyor is unable to do the job, he may be able to recommend another man.

What Does a Survey Cost?

Many landowners have encouraged surveyors to make inexpensive surveys. The sin of these poor surveys is now being felt by present owners in terms of seemingly high costs for resurveys. The cost of a new survey cannot be determined until the records can be examined and the lines checked on the ground. It is possible only to give general guide lines.

Often 1/3 of the cost is studying records, 1/3 is in the field and 1/3 is preparing records of the work done. A good compassman and one good helper can run one mile of random line in 8 hours if the foliage is not too thick. A mile of random line can be offset to the true line, marked and painted by three men in one day. So a resurvey of a small tract might require 3 to 5 days of the surveyor's time. The cost per mile of surveying, marking and painting, plus remonumenting corners, would likely be at least \$175-\$200 per mile. If a transit is used the cost will be higher. To summarize:

1. Cost of a specific job is difficult to estimate.
2. The cost will be higher than the client thinks it should be.
3. Spread over the time the survey should last, the cost is reasonable. A good survey is a sound investment for the future.

Misconceptions

Many people believe a boundary line can be relocated simply by setting up an instrument on a point on (or near) a corner and running the deed call bearing and distance, or by making some "rule of thumb" allowance for needle variation such as 1 degree for every 20 years. A survey made in this way without support of evidence is wrong and a landowner is due for disappointment, boundary arguments, and costly lawsuits sooner or later. He may be passing trouble on to his children or his successors.

Authority of the Surveyor

Every surveyor's authority is derived from his ability, his reputation, and the respect in which he is held. He has no judicial authority. Neither client nor adjoiner is bound by his survey, although both may accept it. A surveyor frequently helps settle boundary disputes by acting as a middleman between adjoiners. When a Court orders a survey, the Court decides where the line is to be.

Deuteronomy 19:14

"Thou shalt not remove thy neighbor's landmark which they in old times have set."

Chapter 7

COST OF SURVEYS

The question most frequently asked the surveyor is: "How much will it cost?" This chapter gives the author's detailed costs of land surveys, mostly in West Virginia.

Costs are based on $4.00 per hour for the party chief and $1.25 per hour for helpers. They must be adjusted to conform to current wage rates. One or two helpers were used on the larger wooded jobs, whereas only one and sometimes none was used on the city and small urban jobs. Costs do not include automobiles travel time or expense, lunch time, cost of pipes, or printing of maps, unless otherwise noted. They do include unpacking at the car, walking to and from the job, packing, establishing markers, and office work. In other words, these costs are mainly for effective job time. Most surveys were closed traverses.

Rural Boundary Surveys

Group A includes typically rural wooded areas where random compass lines were run and offset to the true line, marked and painted.

Group B includes small rural-type tracts too large to be called urban lots. Many lines required no marking because there was no timber present, good fences existed, or corners were intervisible. In some cases transit stations were made in the interior and corners sighted on by triangulation.

GROUP A		
Poles	*Acres*	*Cost*
3022	1015	$1413
597	114	188
340	49	132
2016	1150	619
3044	730	1694
180	20	126
163	10	132
615	107	280
1010	347	711
97	4	88
836	209	425
11920	3755	$5808

GROUP B		
Poles	*Acres*	*Cost*
715	53	$265
247	12	115
229	20	77
261	24	83
177	8	29
189	23	60
152	4	70
103	2	112
460	45	94
479	52	206
298	25	68
422	47	517
310	36	246
218	12	190
295	30	99
512	26	226
209	7	74
5276	425	$2531

	GROUP A	GROUP B
Cost per pole	$ 0.487	$ 0.480
Cost per acre	1.547	5.955
Cost per mile	155.84	153.60
Average size of job	341.4 ac.	25.00 ac.

The 2016 and 3044 jobs in Group A show the extremes that may be found in mountain surveys.

Job 2016 had no unusual problems; there was little brush to be cut, little corner verification needed; labor and cost were $0.307 per pole, $0.54 per acre and $98.20 per mile.

Job 3044 consisted of 10 deed parcels in three consolidated tracts and had four boundary disputes; brush was heavier and more typical; much corner verification and adjoiner deed checking was involved; labor cost was $0.556 per pole, $2.32 per acre and $177.90 per mile. A more detailed study of Job 3044 follows:

ITEM	*RANDOM LINES*	*FINAL LINES*	*OTHER COSTS*	*TOTAL COSTS*
	(12.375 miles)	*(9.51 miles)*		
Surveyor	$308	$300	$806	$1414
Helpers	143	133	4	280
Car mileage	45	41	52	138
Paint (6 gal.)		18		18
	$496	$492	$862	$1850
Cost per mile—$40.05 (12.4 mi.)				
Cost per mile—$52.10 (9.51 mi.)		$51.75	$90.65	$194.50
Cost per acre				$2.53
Poles per man-hour	18.8	15.3		

The error of closure for Job 3044 (three traverses) was: 1 in 1830, 1 in 1080, and 1 in 850—better than might be expected for a 4″-needle compass survey. 22 corners out of 47 were identified. The OTHER COSTS were unusually high and are broken down about this way:

¼ presurvey (abstracting and plotting)
½ during survey (investigations, line disputes, etc.)
¼ after survey (descriptions, calculations, plats)

One West Virginia company reports these costs for rural surveys:

MILES	*DATE*	*METHOD*	*COST PER MILE*
61.5	11/64 to 1/66	transit	$234
22.4	11/66 to 1/67	compass	220
7.4	January 1967	compass	172

Costs include travel time, mileage, random lines, offsets, marking, painting, posting, markers, paint and other supplies, lunch and office time, etc. The company now uses compass for most of its rural surveys.

The following report analyzes an open-sight compass survey of 1406 acres of rough wooded mountain land in Upshur County West Virginia. The land is typical of millions of acres in Appalachia. A plat of the survey plus a typical portion of the description and part of the report of survey are shown in the Plat and Description Chapters.

The survey was made at a total cost of $1657 with an error of closure of 1 in 3480. The cost per acre was $1.18 and the cost per mile of actual boundary was $233.79. Below is a breakdown of costs and related data:

	Surveyor hours	*At $5 per hr.*	*Helper hours*	*At $1½ per hr.*	*Miles @ 8¢*	*Misc.*
Record search, office work (plats, descriptions, reports, etc.)	58	$290			$5	$14
Random lines—field	124	620	114	$172	107	4
Offset, mark and paint	40	200	133	200	34	11
TOTALS	222	$1110	247	$372	$146	$29

The survey was made with foliage both on and off the trees. Surveyor's random hours exceeded helper hours because helper lived close the job, and surveyor-hours included travel one way.

	Feet	*Poles*	*Chains*	*Miles*
Total distance run random	45672	2768	692	8.605
Actual boundary perimeter	37422	2268	567	7.088

Number of compass setups 452
with full tape (or full ones possible) 260
obstacles required less than 2 chains . . . . 168 428 = 1.62 ch. per shot.
recorded offsets around obstacles 24

Out of a total of 17 major corners, 13 markers called for by previous surveys were found, leaving 4 to be located by intersection of deed calls. Streams between some markers provided additional control. With an accuracy within Fourth Order, no problems are foreseen with this resurvey.

These data indicate that rural compass surveys of woodland, including marking and painting, should cost $200 to $250 per mile in the early 1970's. The client "saves" about $50 per mile by furnishing helpers.

Circuit Court records show interesting costs of old surveys in this area. A 1917 survey figures out at $30.00 per mile. A 1931 surveyor who "went upon the land" in February and partitioned 360 acres of rough mountain land into six parcels spent 26¢ per acre. A 1939 survey for ejectment cost $42 per mile.

City Lot Surveys

Below is a summary of 59 surveys of town or city lots and urban-type parcels. They usually required the use of a transit, located either on the corners or on an interior point from which corners were sighted. They are classified into three groups: easy, average, and difficult:

An *easy* survey is one where monuments exist at two or more corners, and subdivision markers and street centers check satisfactorily.

A *difficult* survey is one where no lot markers exist, where there are no markers in the immediate block, and street lines are vague.

An *average* survey is somewhere between these two extremes.

Here again, as in *Rural* surveys, the cost is for *time* of the surveyor and helpers, including setting and referencing of markers. It does not include mileage, supplies, map printing, etc.

	EASY	*AVERAGE*	*DIFFICULT*	*TOTAL*
Number of lots	16	36	7	59
Total cost	$524	$1935	$717	$3176
Perimeter (feet)	16593	33116	5185	54894
Acres	20.08	48.62	4.53	73.23
Cost per 100 feet	$3.16	$5.84	$13.83	$5.79
Cost per acre	$26.10	$39.80	$158.28	$43.37
Cost per job	$32.75	$53.57	$102.43	$53.83
Cost range per lot	$20-60	$27-102	$69-144	$20-144
Average size of lot	1.255	1.351	0.647	1.241

Rural Surveys That Were Not Closed Traverses

This group includes 11 compass traverses totalling 3129 poles—surveyed, marked, and painted. Labor cost was $1716. Cost per pole was $0.50. About half the jobs involved disputed lines and vague corners.

One traverse showed that a capable crew of 3 men ran a 320-pole line in level to gently rolling woods in late October, with leaves partly off, in 6.5 hours. One corner was found and referenced; the random stakes were offset (four feet the maximum); the true lines were marked and painted, in 4.1 hours. Cost was 10.6 x $4.00 or $42.40 plus 21.2 x $1.50 or $31.80; total $74.20. One gallon of paint was used.

Some 1973 Costs

Two tracts were surveyed, monumented, marked and painted as follows:

23 acres of woods divided into two parcels; easy survey—$7.60 per acre.

6.3 acres mostly wooded—$19.75 per acre.

A random line around a 13½ acre tract, ½ wooded cost $13.60 per acre. Several line disputes were involved. The additional cost to offset, monument and mark the lines was estimated at $4.00 Per acre. Total $17.60 per acre.

Two lines each nearly a mile long and 70% wooded, were surveyed, offset, marked, monumented and painted in 54 on-the-job man-hours.

Present Survey Charges

The author currently charges $5.00 per hour, whether it be for record room work, field work or office work. The rate would be about $8.00 if the author was dependent upon surveying for a living. Travel time is charged one way unless the trip is less than a full day in which case both ways are included. Mileage is charged both ways at 10¢ per mile. Other items are charged at actual cost.

Eleven local members of a survey chapter were recently canvassed about their rates for their spare-time survey work. The average was $5.50 per hour and the average mileage charge was 10¢.

According to figures in the *West Virginia Surveyor* for January-February 1973, 35 questionnaires on rates were returned out of about 200 sent out. 63% were from part-time surveyors and 37% from full-time persons. These average charges were reported:

2-man party	–	$14.00	Per hour
3-man party	–	18.00	per hour
record search	–	7.00	per hour
land surveyor	–	9.00	per hour
mileage	–	11¢	per mile

The author will not attempt further discourse on current wage rates and fee schedules. The figures are so variable that it would require much space to analyze them, and rates are going up steadily. The American Congress on Surveying and Mapping assembled a Manual in 1970 showing fees reported by many state associations.

Surveyor Sued

A pier and abutment footing built out of line for a four-lane bridge near Weirton, West Virginia has resulted in a $160,000 lawsuit. Last week the prime contractor, Ryan Construction Company, Janesville, Wisconsin, brought suit against the surveying subcontractor, Construction Layout, Inc. (CLI), Columbus, Ohio, and its surety company.—*Engineering News Record*, March 25, 1971

Chapter 8

THE ATTORNEY–SURVEYOR RELATIONSHIP

Attorneys have played an important part in the development of this Nation, particularly in the field of real estate. Attorneys are in general relied upon to investigate the quality of title to property and to write deeds conveying title to land. Law schools, however, tend to omit many of the legal principles of land survey work.

The surveyor was often looked upon as a technician, relegated to a secondary role, and was certainly not considered a professional. This was the rule in spite of many great men in our nation's past having done surveying. Surveyors were often the pawn of the landowner and politician.

When engineers took over some of the surveying work, the image of the surveyor was elevated in the public mind. However, engineering schools, that once had several courses in land surveying, now generally have few surveying curricula, and give little or nothing on boundary law. In recent years, engineers have entered new and complex scientific fields, and land surveying is being done by less professional hands. In West Virginia most of the rural land surveys are being made by persons who are not engineers.

Neither the attorney nor the surveyor knows enough about the work and problems of the other. These two closely-related practitioners ought to work more closely together. There are four distinct areas in which surveyors should work with lawyers:

1. Land surveys for real estate transactions.
2. As an expert witness in lawsuits.
3. As a source of information about local land.
4. Interpreting land descriptions.

Many attorneys do not understand the requirements of a good survey; they think of surveyors only as "measurers". They are prone to prepare legal descriptions without proper knowledge of the facts or without consulting a surveyor. Attorneys often assemble calls from deeds (surveys) made many years apart, for a new deed, without realizing that the bearings are not compatible. Surveyors sometimes write deeds for conveyances. Attorneys and surveyors should stay within their own fields, and know something of the others' requirements, and cooperate when the need arises.

State and local associations of attorneys and surveyors have an excellent opportunity to create liaison committees and to appear as speakers on each other's programs. Working together, these two groups can do much to educate the public to want proper surveys and upgrade the practice of surveying, so that the surveyor will be looked upon as a capable technician, and eventually as a professional.

Chapter 9

INSTRUMENTS, EQUIPMENT, TOOLS AND SUPPLIES

A competent surveyor must have good equipment. This chapter lists equipment and some sources of supply. Many items are essential; some are helpful but may be omitted; a few can be homemade. A list of tools to be taken to the field should be made on a card that can go in the front of the field book. It can be checked before a trip and before starting home so no item will be left in the woods or in pockets of helpers. A compass surveyor should invest at least $1000 for new equipment, exclusive of his automobile or truck. Sometimes the equipment of an inactive or deceased surveyor can be acquired. Sources of supply include hardware and office-supply stores, but most items peculiar to surveying can be obtained from these companies:

Warren-Knight Company, 2045 Bennett Road, Philadelphia, PA 19116.

Forestry Suppliers, Inc., P.O. Box 8397, Jackson, MS 39204.

The Ben Meadows Company, Box 8377, Station F, Atlanta, GA 30306.

The H. T. Hall Company, Inc., P. O. Box 1549, Charleston, WV 25326.

Buff & Buff Mfg. Co., 329 Lamartine St., Jamaica Plain, MA 02130.

Field Equipment—Major Items

Axe, hand, 1-1½ lbs. with sheath.

Ball-joint with tangent screw (for Warren-Knight Compass) No. 25-430 T.

Boots or 6″ shoes with Vibram soles (Bean's Hunting Shoe No. 3175.)

Canteen, polyethylene, 1 qt.

Case for compass, leather (Warren-Knight No. 130 C).

Clinometer, Suunto, Degree-percent, with case.

Compass, open-sight, in wood box (Warren-Knight No. 130).

Compass, hand, Silva Ranger, in quadrants, with case.

Field book, looseleaf.

First aid kit, snakebite kit, insect repellant.

Flagging, orange glo or high visibility or fluorescent vinyl.

Magnifier, non-magnetic, 2 or 4 power.

Markers, aluminum caps for 5/8″ reinforcing rod (Allied Surveyor Supplies Mfg. Co., Box 4309, Tucson, AZ 85717). Use with 1/8″ number and letter dies, anvil and driving cap.

Pencil, 3H or 4H, non-magnetic.

Pruner, shrub, Stanley PR19, heavier types for lineman.

Rod or range pole, developed by author especially for opensight compasses; aluminum, two sections, 2″OD, 7 to 8 ft. long. (Buff).

Slope reduction tables, by ½° from 3° to 40°, for distances of 1.000, 1.250, 1.500, 1.750, and 2.000, on 4.6″ x 7.2″ cards, (from author for 25¢ in stamps or cash). See pages 122 and 123.

Tape, 200-link, band chain, no reel or trailer (K & E Warrior).

Tape (alternate choice), 200-foot utility fiber glass (Lietz).

Tape, steel, engineers, 50 feet, in 10ths and 100ths.

Tape repair kit, Eureka to fit tapes (repair instructions later).

Tripod, aluminum, telescopic (Warren-Knight No. 130 T).

Vest, Filson blaze orange cruiser sizes 36 to 50, for compassman and rodman.

Field Equipment—Miscellaneous

Aerial photo carrier (flexible), brief case, bump cap, cold chisel, marking crayon, file, garden trowel, knapsack, knife, nails (roofing and 60d), bottle and jar caps, pipe 1″ ID or larger, paint scraper 2½″, plumb bobs, targets and string, sandwich boxes, bow saw 24″, trench shovel, watch, whetstone, thermos bottle.

Office Equipment—Major Items

Calculator.

Copier, photo.

Drafting machine, 1′ vernier, adjustable orientation, with 12″ rulers 10-20-30-40-50 scales.

File drawers (for 9 x 12 size).

Lamp, drafting table.

Planimeter.

Stereoscope, pocket, with case.

Table, drafting, 3 x 5.

Tables, Standard Field (BLM).

Tables, natural cosines and sines (Calders).

Table, tracing, Portatrace or homemade.

Typewriter, portable electric.

Office Equipment—Miscellaneous

Curve (French or railroad), drafting papers and films, manilla folders 9″ x 12″ with 5-cut tabs, ruling pen, drop-center pen, pencils 2H to 8H, colored pencils, protractor (Plotractor), 6″ flat scales 10-20-40-50, pencil sharpener, erasing shield, stapler, drafting stool, drafting tape ¾″ or wider, templates with circles, triangles, etc., mailing tubes, envelopes, magnetic tables, triangles (4″ 30-60, 10″ 45-90), T-square or straightedge (36″).

Chapter 10

THE SIPE-SUMNER VERNIER COMPASS

History of the Compass

In 1965 the author published a book about land surveying. In 1966 the late Gerald R. Hyde, a land surveyor and forester in Bedford, New Hampshire, bought several of these books. Hyde sent a review of the book to the American Congress on Surveying and Mapping. The review was published in the December 1966 ACSM Journal Vol. XXVI No. 4. In January 1967 the author received a letter from L. Bruce Sumner, a U.S. Government surveyor in San Francisco, California, stating that he had read the review and was interested in compass surveying.

This led the author to express the need for a new open-sight compass, in which Sumner concurred. Following many communications between the parties, Sumner by late 1967 had developed a set of specifications.

Hyde meantime had broached the idea of a new compass to the Buff and Buff Manufacturing Company and in November 1967 Sumner sent preliminary specifications. Two models were completed in 1969 by the Buff Co. but they were unable to proceed with the project.

In March 1971 Sumner visited the Warren-Knight Company in Philadelphia. President John H. Warren, Jr. and Vice-President Harry W. Fatkin were interested and soon agreed to develop and manufacture the compass. By August 1971 the design had been finalized and the first production prototype #17841 sent to the author and the second #17842 to Sumner. Production had started.

As of now several hundred compasses are in nationwide use. Warren-Knight also make, especially designed for the Sipe-Sumner Pattern Compass, a tripod and a Ball-Joint with Tangent Screw. The new compass outfit, designed for serious compass surveys, embodies the best features of previous compasses.

General Description

The compass has a 5″ needle, a one-minute vernier, and is graduated in quadrants to half-degrees. Adjusting knobs are located to avoid catching on obstacles and can't be lost, yet are in handy positions for use. The compass is designed for easy operation, adjustment and repair, and is considered a life-time instrument. Each instrument is serially numbered. Weight of the compass alone is 2¾ lbs. It is 6½″ wide, 7½″ long, and 2″ high with sights folded. Following are other specific features:

Vernier

A one-minute vernier operates smoothly with precise drive and control knob, allowing easy set-off of declination (to 40° or more) and variation (secular and daily). Index error is reduced to a minimum. Handy and positive clamping knob secures any setting.

Sight Vanes

Rear slit and front hair type, 6½″ long, easy lifting, snug-folding, usable on slopes up to 45°; 6.9″ distance between sight vanes allows more precise sighting. Vanes adjustable for verticality; 90° offset sights.

Needle

Length 5″; neither too sensitive nor too slow; hardened steel pivot is removable for sharpening or replacement. Tips in same plane as synthetic sapphire pivot suspension; will remain steady despite residual lateral movement of needle. Counterweight adjusts easily for the latitude of the user. Tips colored.

Graduations

Cardinal index marks on floorplate rotate with graduated circle and helps prevent blunders in reading bearings near cardinal points.

Level

Easy-reading bulls-eye level of good size and location for quick levelling. Adjustable without removing.

Cover

Tempered glass; leakproof and easily removable.

Housing

Lightweight strong aluminum. Ventilation plugs to relieve condensation.

Ball-Joint and Socket

New design large-diameter ball gives smooth operation and stability. Socket fits telescopic tripods now in use.

Case

Handsome wood case with soft inside cushions. Special leather case available.

Capability and Performance

The final test of the value of an instrument is based on its performance and capabilities. How accurately can the angles be read? What angular error of closure can be obtained? How easy or difficult is it to use? Is it adaptable to a variety of conditions? Is it easy to adjust and repair? Is it well made?

The end product of a practitioner depends not only on the equipment used, but on the ability and care used by the operator and his helpers, and the conditions under which the work is being performed.

A poorly trained or careless operator will obtain poor results from the best equipment. Difficult terrain will make it hard to obtain results more easily attained on favorable terrain. The **Sipe-Sumner Vernier Compass** in the hands of a poorly trained observer could give mediocre results; a proficient user will get excellent results. What results can be expected with the new compass?

The observer must first become familiar with every detail of the instrument and correct operating procedures. It is like learning to drive a new car, with different controls at different places than on the old vehicle. Experience is a good teacher.

How to Use the Vernier

To read the bearing of a line to the nearest minute, release the vernier clamp and set the vernier at zero. Sight approximately on the rod, line, station or object and clamp the spindle of the ball-joint. With the tangent screw rotate the compass precisely to the object. Read the bearing of the north end of the needle to some fraction of the 30 minutes involved, making mental note thereof. Now with the large vernier control knob under the south vane, rotate the outer ring so that the needle points to the nearest lower graduation. For example, if the first reading looks like N 54-10 E, rotate the vernier knob so that N 54 E is opposite the needle tip. (The graduated limb on the far side of the compass will move in the direction the thumb moves if the thumb is on the observer's side of the knob). Now look at the vernier and read the matching line in the direction in which zero of the inner vernier scale has departed from the zero of the outer vernier scale. The lines that match should be in the vicinity of 10, if the estimate was well made.

To set a predetermined bearing, set the odd minutes on the vernier as follows:

NE and SW quadrants: inner vernier zero should go to left of the outer zero.

NW and SE quadrants: inner vernier zero should go to the right of the outer zero.

Now release the spindle clamp, rotate the compass by hand until the next lower graduation mark is opposite the north tip of the needle, and tighten the spindle clamp. Rotate compass with the tangent screw as needed. For example, to sight on S 57-43 E, set 13 minutes on the right side of the inner vernier to match a line on the outer vernier ring; then the outer ring should be adjusted with the tangent screw until the needle reads S 57-30 E. The sights are now on the predetermined bearing.

To set off declination, rotate the graduated limb *counterclockwise* for a *west* declination, and *clockwise* for an *east* declination. Setting off declination is seldom advised.

Field Testing the New Compass

The following tests should be made to test the compass and develope proficiency.

1. Set hub and tack in a flat area. Set 1″ x 36″ dowels about every 45 degrees 40-50 feet away from hub, like wheel spokes. Set compass with plumbbob over tack with spindle vertical (use hand level on tripod head). Read bearings to all dowel tops, using vernier and magnifier. Avoid parallax; watch for magnetism in eyeglasses—stand in same position (at south end of needle) for all observations. Take all readings within 30 minutes* to minimize daily variation. Then set transit over tack, orient on any dowel at zero and read azimuths to each dowel top. Calculate angles between dowels by compass and transit and compare. All must add to 360. When compass angles vary from transit angles by not over 2-3 minutes, compass is in adjustment and observer is proficient.

2. In a flat field free of magnetic disturbance, project a line, setting up compass every 100 feet or every 2 chains. Use nails and bottle caps for markers. Run at least ¼ mile. Set transit at one end station and record divergence of each marker from a line to far end. They should be within 0.2-0.3 foot divergence.

3. Set 5 stakes about 150 feet apart in a flat field so as to make a 5-sided figure. Measure distances between stake tops to nearest 0.01 foot. Set compass over center of each stake by plumbbob, with spindle vertical. Read bearings forward and backward using vernier. Maximum deviation of foresights and backsights should not be over 3′ each; sum of interior angles within 10′ of formula; error of closure of fore or backsights from 1 in 5000 to 1 in 10000.

Tip

When turning the chain, grab it in the loop and walk with it and let the end come to you, then go.—by Glenn A. Eslinger, in W.VA. SURVEYOR

*of time

Chapter 11

TESTING AND ADJUSTING THE COMPASS

by
L. Bruce Sumner
Cadastral Surveyor
U. S. Forest Service, California Region

It is unfortunate that most compasses in use (including those on transits) are in unserviceable condition. Although it may have been reliable when new, the average compass has been subjected to misuse, knocks and wear. The surveyor must, therefore, be able to determine the condition of his compass. When deficiencies are found, he should know how they can be corrected.

Although the open-sight compass is a theoretically simple instrument, the adjustment of its various components is most exacting. If any part is out of order, restoring it to proper condition requires great care and a delicate touch. This is especially true when dealing with the needle or its pivot. The tests and corrections to be made before the compass can be considered fit for duty are described below in sequence.

The Level

After levelling, the level(s) should remain centered when the instrument is turned in any direction. This requirement is essential to the proper functioning of the sights and needle. There are two causes for failure of this test: levels out of adjustment, or a faulty mounting spindle. If it is the spindle, either new parts or remachining will be necessary.

The level of a compass with two tubular level vials is normally adjustable, either by opposing screws or by insertion of thin shims. The principle of reversion is used, just as in adjusting the levels on a transit. Half the deviation after rotating the compass 180 degrees, is corrected by relevelling the compass, and the other half by adjusting the levels themselves. This is done for one vial at a time.

The same principle is followed for those compasses using a single "bulls-eye" level. However, some of these have no provision for adjustment. For such non-adjustable levels, determine a position where the bubble does not move when the compass is rotated 360°. This position can then be marked with a dot or fine circle of red paint and the factory marking ignored.

Verticality of Sights

Both sighting slits, or slit and hair, must be truly vertical when the compass is level or the projected line of sight will be erratic, and errors will be introduced into the observed bearings.

The test is made by sighting on a plumb line after carefully levelling the compass. First, take a piece of masking tape long enough to cover the slit in the south vane and make a hole about 3/32″ in diameter in its center. Lightly stick the strip of tape to the south vane so that the view through the slit is obtained through the peephole. Because the eye is held in vertical position by the hole while observing the relation of the north sight slit or hair to the plumb line, the lack of verticality in the south sight has no effect on testing the north sight. The south sight is tested by sticking the tape with peephole to the north sight and observing the relation of the south slit to the plumb line. If the north sight has a vertical hair, the peephole can be offset to the side slightly if the hair interferes. It is normally so close to the eye as to be out of focus and almost invisible.

On those compasses with sights mounted on a horizontal plate or extension, the adjustment is simple and positive. Thin pieces of brass (never steel) shim stock (or just paper in emergency) are inserted under the edge of the sight base to move the sight to a truly vertical position. There should be no sideways play in the hinge joints of a compass with folding sights.

Many compasses now in use have the sight bases screwed to the *side* of the compass shell. These are held in place by friction and in some cases also by the wedging action of counter-sunk screw heads. Adjusting this type of sight is a trial of the individual's patience and skill. It sometimes requires reshaping of the holes in the sight base, and the sight must be held in exact position while the screws are tightened. This adjustment is made more secure by coating the mating surfaces with a slow-setting varnish and leaving it undisturbed until the varnish is hard.

Declination Offset

The crude declination-variation indicators on compasses usually cannot be set closer than the nearest 5 or 10 minutes. Tests described elsewhere in this book show that compasses (even of good transits) may vary over one degree if compared on the same line. Therefore, the wise surveyor will test his compass and determine its index error. One method for doing this is described in other chapters.

Keep in mind that any adjustments made to the needle, pivot or sight bases will require that the index error of the compass be redetermined. A drop of varnish or cement will secure the offset mechanism on those compasses without a positive clamp.

The Needle and Pivot

The Pivot Point. This is the greatest single source of poor directional results in compass surveys. A dulled point results in a sluggish or sticky needle. It has been this writer's observation that many persons who loudly proclaim the "unreliability" of the compass also treat one in such a manner that the pivot point is quickly ruined. It should go without saying that a compass should never be carried or subjected to jolts and vibration when the needle is resting on the pivot. When releasing the needle, it should be lowered gently onto the pivot point, not let down suddenly.

Tests.

1. Remove all ferromagnetic objects such as knives, mechanical pencils, wrist watches, etc., from the person. Obtain a 12 oz. "tin" can (Sheet steel). Set the compass level on a tripod or workbench with the can 28″ distant horizontally and magnetic north from the compass. This can be done inside the home or shop so long as all ferromagnetic objects remain static in position. Nearby radios or television sets should be kept on or off, because certain components produce a magnetic field.

2. Rotate the compass so that the north tip of the needle is exactly on a mark when freed and at rest. When the can is moved *one foot* to east or west, the needle should move slightly off the mark; this test will indicate *sensitivity*. An exceptionally good compass will react to a 12-oz. can 36″ from the north tip of the needle.

3. With the needle free and at rest on a mark, distract it by moving the can near one side and then replacing it in its previous position. The needle should come to rest exactly on the mark *ten* times out of *ten;* This test will indicate *reliability*.

The average compass in general use will fail both of these tests in a most convincing manner.

Correction. Disassemble the compass enough to remove the needle and pivot point. The latter is usually removed by unscrewing from above or below. If stubborn, apply a little kerosene or gun solvent to soften rust and corrosion and let stand for a few hours before removing. Examine the point under magnification. It will often be found to be so blunted that it resembles a crochet needle. It should have such a fine point as to appear absolutely sharp, even under 10-power magnification.

Before the pivot can be properly sharpened, it must first be clamped into some sort of holder. A good holder for most pivots can be made from an ordinary "lead" pencil stub. Split a pencil of appropriate lead diameter lengthwise along the glued seam and remove the lead. The pivot shank is inserted in the open end, and clamping action is supplied by tightly wound rubber bands. Once in its "handle", the pivot may be carefully sharpened on a fine stone such as a razor or Arkansas stone. The common stones sold for sharpening knives are much too coarse. The pivot is rotated while stroking on the stone to produce a uniform conical point. It should be held at approximately 30° from the stone to produce a point with about 60° included angle. (See diagram.) Keep the point angled slightly in the direction of stroking to avoid building up a burr on the tip. Bear in mind that sharpening tends to shorten the pivot, thus lowering the needle tips.

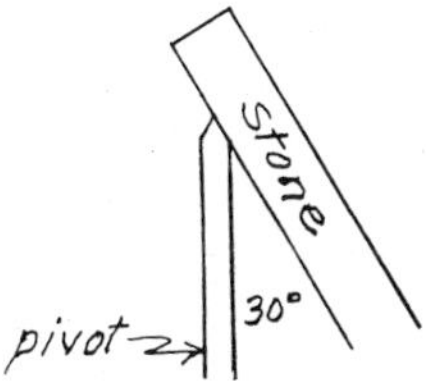

The needle center-bearing (usually agate) is cleaned by rotating a round toothpick wrapped with lint-free tissue or cloth in the socket. The final cleaning is done with tissue or cloth dampened with lighter fluid or alcohol. All oily residue must also be removed from the pivot point if good needle-action is expected.

When reassembling, place the needle on the raised lifter rather than directly on the pivot point, and then lower onto the pivot. Failure to follow this procedure has spoiled many sharp pivots. Before all adjustments are complete, the needle may have to be removed and replaced a number of times. A needle with a good pivot is slow to settle, so release it when pointing northward to minimize swing.

Horizontal Alignment. A free needle properly made and in good adjustment will have the tips level with, or very slightly above, the upper surface of the graduated circle when the compass is level. If much above or at all below, the resulting parallax can cause inaccurate observations. If properly constructed, a line through the upper surface of the tips will be coincident with the junction of pivot and bearing. This will result in no relative movement between the needle tips and the graduated circle although the needle is still vibrating at right angle to its long axis. This residual vibration or "shimmy" for sometime after the needle stops swinging can be noted by watching the reflected light from its upper surface, or that of the pivot jewel. The presence of this phenomenon indicates a good pivot.

Some sources recommend bending the ends of the needle upward (or downward) to make the tips run level with the graduated circle. Any such bending will destroy the alignment between tips and pivot in the horizontal plane. The proper way to accomplish this objective is to raise or lower the pivot. When the pivot point is too low from repeated sharpening (or too high for some reason), it is best to balance the needle so that the top of the north tip is level with, or about 1/64″ above the circle and forget about the south end. Proper compass practice dictates exclusive use of the north end of the needle, observed while standing at the south end. A compass with a pivot of adjustable height would be superior to the current offerings.

Once the needle is balanced by the sliding counterweight on the south end, it need not be changed while working in the same general area (normally within a hundred-mile radius). However, there may be occasions where the needle is found to be very badly out of balance for no apparent reason. This is nearly always the result of a static electric charge on the cover glass. Such a static charge can develop almost spontaneously during periods of low humidity. To dissipate the static, merely moisten a finger tip and rub it around the cover glass about three times, and the needle will assume its proper attitude.

Vertical Alignment. The essential element of alignment in the vertical plane is for the graduated circle to be concentric with the rotational point (pivot assembly) of the needle. If this criterion is not met, the compass will have a built-in *variable* bearing error depending on the direction sighted. It is analogous to eccentricity in the plate of a transit.

Test. The north and south tips of the needle should cut opposite graduations when the compass is pointed in any direction. If they do not, corrective measures must be made. The south tip of the needle has value, but for test and adjustment purposes only.

Correction. Rotate the compass with needle free until the position of greatest deviation between the tips is found, and note the position. Place the north tip on the nearest mark and observe the direction of error of the south tip. Remove the needle and gently bend the pivot at right angles to where the long axis of the needle was located and opposite to the error observed in the south tip. This is a trial and error proposition, so proceed slowly, making adjustments a little at a time. Replace the needle and check, repeating as needed. Holding ferromagnetic metal object near the north end of the needle can help to make the needle come to rest more rapidly; proper manipulation will require a little practice.

Once the needle is made to cut opposite marks, rotate the compass 180° and see if it still cuts opposite marks. If it does not, the tips and suspension point are not in a straight line. Correct half the error by delicately bending the needle in a horizontal plane and the other half by bending the pivot. Once this is done, the tips and pivot are aligned and all further adjustments are made to the pivot only.

Next test the compass in the same manner with needle at 90° from the first test, making corrections by bending only the pivot. This can be done with needlenose pliers, grasping pivot at least 1/8 inch below the tip. To refine the adjustment, test and correct as needed at 45° each way from the original setting.

The final result should be a needle that cuts exactly opposite marks when rotated to any position. This requires a lot of work and many repetitions, but it is necessary for the upper limits of compass accuracy. However, if the deviation between needle tips is kept to less than $1/4^{\circ}$, it can be considered a serviceable instrument for ordinary work.

Conclusion

This is not meant to be the whole story on testing and caring for the magnetic compass. For those so inclined, it will provide a guide for testing and mechanical adjustment. For those who have the work done commercially, it will allow double checking what was done.

The value of knowing *personally* what your surveying instrument will or will not do, cannot be over-emphasized. Having restored and used various instruments which were made over a span of more than 125 years, the writer is a firm believer in this principle. A good compass is a delicate scientific instrument which commands respect. Treat it right and it will serve you well.

An Honest Mistake

Somerville et al. vs. Jacobs. Wood County, West Virginia. Civil Action No. 9176. Where through a reasonable mistake of fact and in good faith one erects a building upon the land of another, with reason to believe that the land thus improved is that of the improver, the landowner must either pay the improver the amount by which the value of his land has been thereby improved, or sell the land upon which the building has been erected to the improver at its fair market value immediately prior to the erection of the building.

Brooks Bill Signed into Law (from W.VA. SURVEYOR)

After a five-year fight, design professionals have won the biggest legislative battle they had ever undertaken. President Nixon has signed the Brooks bill into law, making it official government policy to select architects and engineers on the basis of competency rather than price competition.

The General Accounting Office had tried to persuade Nixon that the bill was either unwise or unnecessary—at least until the Commission on Government Procurement reports its findings, sometime next month. Later, however, GAO modified its stand. The Justice Department's antitrust division also tried to get the bill vetoed.

The new bill makes almost no changes in existing government practice, which has been to select A-Es on competency anyway. But the design professionals feel it is highly important to establish the principle and to have it written into law.

Chapter 12

HOW STRAIGHT A LINE CAN BE RUN WITH AN OPEN-SIGHT COMPASS?

A survey was made to see how straight a line could be run with an open sight compass where brush, rough ground and vertical angles were not factors. The effect of daily variation was to be studied. In other words, a "straight" line was to be projected, and the deviation from said line caused by the variation of the needle during the survey was to be evaluated.

On February 27, 1968, between 2:00 PM and 3:00 PM a line was surveyed N 89 W a distance of 46 chains through a level grass field in Florida. An open-sight compass with 4″ needle, tripod and 2-chain tape were used. The wind was high enough to require legs of the aluminum tripod to be firmed in the ground at each setup. The wind made the tripod vibrate enough so that it was not necessary to tap the tripod to insure that the needle was swinging freely. A nail and bottle cap was set every 2 chains, making 23 compass stations and observations. No back-sights were observed. No plumb bob was used; the compass was set over each station by eye; the error was about 0.05 foot per station as determined by actual field test. No local magnetic attraction was apparent.

The second phase was to set a Wild T 16 ED transit at station 0, set plates at 271-00.0 and orient on station 4, which was 4 chains away. A rodman then proceeded to each station, to which an azimuth, estimated to the nearest 0.1′, was taken from station 0. The azimuth from station 0 direct to station 23 was 270-54.0, and this was used as the "base line" from which to figure the divergence of the intervening compass stations.

The table on the next page shows the result of the experiment. The reasoning behind the calculations for deviation is as follows: The U. S. Government variation tables for Cheltenham on February 27 show the needle to be about 4.0′ left of mean at 2 PM, and 3.35′ left at 3 PM (interpolating between February and March figures). Thus the needle has a net movement to the *right* of 0.65′ in that hour. Distributing this 0.65′ through 23 setups shows the change to be about 0.03′ per station. So at the end of the first 2 chains, if no allowance was made for variation, the compass would be 0.0011 ft. to the right of the base line (using the formula: divergence equals 132 ft. x tan of 0.03 minutes). At 4 chains, the compass would be 0.0011 plus 0.0022 equals 0.0033 ft. to the right, and so on until at 46 chains (and one hour later) the compass would be 0.304 ft. to the right of the base line. This, of course, assumes a uniform time lapse between stations.

In other words, if the difference in variation between 2 PM and 3 PM were averaged (i.e., ½ the difference) and used as a correction factor to 46 chains, the result would be about the same at the ending point as if each station were progressively offset.

It will be noted that:

1. During the 60 minutes used to run the line, the daily variation should curve the line about 0.30 ft. to the right of a straight line.

	Calculated Effect of Daily Variation				
Station (chains)	*Shift of Needle (minutes)*	*Divergence Per 2-Chain Shot (feet)*	*Cumulative Divergence To Right (feet)*	*Actual As Run Divergence From Base Line (feet)*	*Net Compass Error of Divergence (feet)*
2	.03	.0011	.0011	.23 R	.23 R
4	.06	.0022	.0033	.10 R	.10 R
6	.09	.0033	.0066	.27 R	.26 R
8	.12	.0044	.0110	.00	.01 L
10	.15	.0055	.0165	.25 R	.23 R
12	.18	.0066	.0231	.44 R	.42 R
14	.21	.0077	.0308	.35 R	.32 R
16	.24	.0088	.0396	.48 R	.44 R
18	.27	.0099	.0495	.34 R	.29 R
20	.30	.0110	.0606	.19 R	.13 R
22	.33	.0121	.0727	.21 R	.14 R
24	.36	.0132	.0859	.14 R	.06 R
26	.39	.0143	.1002	.29 R	.19 R
28	.42	.0154	.1156	.54 R	.42 R
30	.45	.0165	.1321	.23 R	.10 R
32	.48	.0176	.1497	.38 R	.23 R
34	.51	.0187	.1684	.20 R	.03 R
36	.54	.0198	.1882	.40 R	.21 R
38	.57	.0209	.2091	.22 R	.01 R
40	.60	.0220	.2311	.40 R	.17 R
42	.63	.0231	.2542	.25 R	.00
44	.66	.0242	.2784	.09 R	.19 L
46	.69	.0253	.3037	.00	.30 L
				Average per station 0.26 ft. R	0.19 ft. R

2. The maximum actual divergence between the most northerly and southerly station was 0.54 ft.

3. The maximum divergence with the progressive change in needle direction allowed for was 0.44 ft.

4. All actual divergencies were to the right of the base line.

5. Average divergence without allowing for change in needle direc-

tion was 0.26 ft., and with the calculated correction applied was 0.19 ft., nearly all to the right.

6. Without application of the corrections for the changing needle direction due to daily variation, no station was more than 0.54 ft. off the true straight line. The significance of this is better understood by considering a straight line between two points 5620 feet apart; no station would be more than a foot from this line. However, lengthening or shortening the distance between stations would materially affect this deviation. That is, generally the more setups and the shorter the shots, the greater the accuracy produced by the compass. This is of distinct advantage in woodland surveys. When corrections for daily variation are applied, the result is even better.

A Similar Test

Another test of this type, but with the Buff prototype compass, was made in a level field in Florida. A 2500-foot line was run magnetic north. A nail and cap was set every 125 feet, moving the compass forward to each new station. A plumb bob was used. The normal daily variation for the 45 minutes consumed in the test would be about ½ minute to the right; no allowance was made for it. The vernier was set at zero and not used.

The second phase of the test was to set a Wild T16ED transit at the beginning Station A, set the plates at zero, sight on B and release the top plate. Then horizontal angles were observed to each nail and cap in turn (estimated to 0.10 minute). The distance each station was to the right or left of line AB extended, was calculated, with these results:

Station	*Deviation (in feet)*	*Station*	*Deviation (in feet)*	*Station*	*Deviation (in feet)*
A	0.00	H	0.00	O	0.10 R
B	0.00	I	0.09 R	P	0.11 R
C	0.04 R	J	0.03 R	Q	0.28 L
D	0.07 L	K	0.00	R	0.19 R
E	0.10 L	L	0.00	S	0.07 R
F	0.09 L	M	0.00	T	0.07 R
G	0.09 L	N	0.15 R	U	0.08 R

Total Rights–0.60 ft. Total Lefts–0.96 ft.

The position of Station Q 0.39 to the left of line OP if extended provided the greatest crook in the line. Line OPQ was rechecked without finding any significant difference from the first readings. Perhaps there was an underground pipe near Station P.

It should be remembered that in a test of this kind on level land, index errors of the instrument are not a factor; it is a battle between the observer and perfection. The significance of this test is better understood by considering a line 8928 feet long (1.69 miles) on which no station would be more than a foot from the true line. On the basis of the above figures, most stations would be within 4 inches of the line.

If the compass is read with an average error of 3 minutes per observation, the error in one mile can be calculated. Assume that on sloping woodland an average shot is 110 feet long creating about 49 stations per mile. The tangent of 3′ is 0.00087 which creates a divergence in 110 feet of 0.0957 feet (an error of 1 in 1150). The formula for propogation of compensating errors shows that the divergence in one mile is 0.096 times the square foot of 49 = 0.67. Then 0.67 in 5280 is an angular error of 1 in 7880. In the same manner, a 5′ error per station becomes an angular error per mile of 1:4700.

Projecting a Line From Backsights

The need for projecting a line from backsights alone arises when sources of disturbance are encountered, such as garbage dumps, fences, automobiles, etc., and when it is not feasible to offset. The following test was made with the new Warren-Knight Surveying Compass.

A level field free of high grass was selected, and a nail was set at (A). The compass was set up over it using a plumb bob. The tripod head was levelled by eye so the needle pivot would be over the nail. The bulls-eye level was centered carefully. Then an arbitrary direction was selected; two chains were measured and a nail set at (B). (Three range poles were tried. At 132 feet, poles of 7/8″, 1¼″ and 1½″ OD were too small; the front hair obscured the poles. A 2″ OD white pole proved ideal.)

A 2″x2″x18″ wood stake with orange ribbon was then set in line at (A) leaving about 16″ projecting. The compass was moved to (B) and backsighted on the stake at (A). The compassman then moved to the rear of the compass and sighted forward in the normal manner. 2 chains were measured and a nail set at (C). A stake was set at (B) just as at (A). The stake at (A) was pulled up and laid on the ground. The compass was then set up over (C) and the operation repeated. 16 chains (8 stations) were run in this manner.

All stakes were then carefully reerected in line near the nails. A Wild T16 ED transit was then set up over station 16, plates at zero, and sighted on stake (A). The lower plate was clamped and top plate loosened. A sighting was taken on the top of each stake. The table below shows the results:

Station	*Azimuth*	*Calculated Deviation From Line A to I (in feet)*
A	00-00.0	0.000
B	359-59.8	0.056 left
C	00-00.2	0.046 right
D	00-00.4	0.079 right
E	00-00.7	0.106 right
F	00-01.2	0.139 right
G	00-01.6	0.121 right
H	00-01.6	0.061 right
I	(Wild at I)	0.000

The stakes were set in an hour so the daily variation was negligible.

It is obvious that the projection of a line with good accuracy by use of backsights is practicable with the Warren-Knight compass. The probability of errors in each specific operation involved would be not over 0.02 ft. and these would be compensating. The 6.9″ distance between sight vanes and the slit-hair combination allow an adequate sight-radius for the human eye to operate effectively. The test shows why the "picket" method of extending a line can be useful, and how some fairly straight lines could have been run with old compasses that had sight vanes 10-14″ apart. How far can a line safely be extended by projecting backsights? This test indicates that a line one mile long would have its stations off line not over 6 or 7 inches.

One may now ask: why use the needle at all for extending a line; why not just use backsights? The answer is obvious: terrain, vegetation and stake-setting problems provide obstacles that would make such surveys impracticable. But through fields, projection by backsights is satisfactory if the bearing of the first line is good. It is obvious also that backsights may be relied upon to check for unusual needle variations or sticking of the needle.

Chapter 13

COMPASS VS. TRANSIT

In the early years of this country, the link chain and compass were the standard instruments used for surveying land. As land became more valuable and technology developed new equipment, the link chain gave way to steel tape, stadia, lasers, and microwaves; the compass lost its popularity to the transit.

Surveyors, who should have the responsibility to choose the proper instruments for the job at hand, seem to have lost that prerogative. In this regard, the author has found no truer statement than made by Davis and Foote on p. 33 of their 1940 textbook *Surveying:* "There are no set rules for the relative accuracy of different classes of surveys, nor could there be, for the objects and conditions are too many and too complicated. xxx The best surveyor is not the one who is extremely accurate, but the one who can make a survey just accurate enough to serve its purpose without waste of time and money." However, the surveyor today who wants to use the type of instrument which fits his job, will find confusing statements such as these:

Gillespie (*Treatise on Surveying,* 1895): p. 100 "xxx the compass is xxx most commonly used in land surveying in spite of its imperfections and inaccuracies." p. 111 "The compass is deficient in both precision and correctness." p. 112 "xxx the compass is a very valuable instrument xxx and though never precise, and seldom correct, it is generally not very wrong."

Tracy (*Plane Surveying,* 1907) p. 293 "The transit has taken the place of the surveyor's compass to such an extent that now the latter is used for rough work only xxx." p. 298 "If the bearing is correct to five minutes, which is as close as the compass can be read with a reasonable degree of certainty xxx."

McCullough (*Practical Surveying,* 1915) p. 6 "xxx readings cannot be taken closer than ¼ degree with the best made compass." p. 103 "xxx its (the compass) abandonment as an instrument of precision xxx." p. 104 "xxx a well-trained conscientious surveyor, using a well-made modern compass, can do very good work." p. 114 "Closer readings than ¼ degree are only estimates." p. 217 "A good compass is graduated to ¼ degree; by using care, one half this angle may be estimated. xxx the minimum error in angle will be 4 minutes xxx."

Foster and Degroot, (*Surveying,* circa 1936) p. 20 "It is useless to try to read a compass closer than 10 minutes because of the inaccuracy of the instrument."

Debenham, (*Mapmaking,* circa 1936) p. 61 "xxx compass bearings xxx can be relied upon in circumstances where one would not trust a more accurate instrument. xxx the compass xxx deserves a better reputation than it has with those surveyors who regard precision as a fetish to be worshipped at all times."

S. E. Huey, (*ACSM Journal "Survey & Mapping,* Vol. XII No. 3, 1952) p. 296 "with reasonably accurate chaining and careful use of the transit's compass, surveying with a closure of 1/3000 can be made, which is a precision sufficient for most land surveys xxx."

ACSM Journal (*"Surveying & Mapping,* March 1958) p. 72 "In Trinidad xxx the average error of closure of 323 compass traverses run between trigonometric stations about one mile apart is less than 1 in 2500."

Brown & Eldridge (*Evidence and Procedures for Boundary Location,* 1962) p. 203 "The method of observing magnetic bearings for each side of a property has passed into obsolescence xxx compass needle readings can at best be read only to the nearest quarter degree."

DeMoisey (*Forest Surveying,* 1965) pp. 36, 38 "The compass circle usually is graduated in half degrees xxx and bearings may be read to the nearest degree, or estimated to the nearest degree, or estimated to the nearest one half degree. xxx The accuracy ratio is about 1/300."

After reading such contradictory statements, it is understandable how the man who has not used a compass much will likely be critical of its capabilities. Its weaknesses *seem* to stem from these facts:

1. Compasses have accidental index errors built into them.
2. Pivot and sight vanes can get out of adjustment.
3. Different (even the same) observers can read different bearings on a given compass at a specific station (parallax, etc.).

Some of these errors are illustrated by this test conducted by the author: seven men, each with a different transit sitting on true north lines 20′ apart, read the declination of their own instrument. The declinations were: 4-45, 5-00, 5-05, 5-10, 5-15, and 5-30. The declination for this area was calculated by the Government to be 5-11. A similar test made by L. B. Sumner, a California surveyor, on a true north line with six assorted compasses and transits, showed differences in declination of 1°45′.

The *fallacy* of concluding from the above data and for other reasons, that the compass needle is "unreliable", is really a *failure to understand these significant objectives of a boundary line survey:*

1. To establish or reestablish referenced markers at the property corners, located in accord with predominating evidence and the rules of law, considering the rights of adjoining properties, so that the markers may be found in future years by an observer in the vicinity.

2. To connect said markers by lines of reasonable or specified straightness, marked so they are visible to an observer in the area.
3. To prepare a plat and description of the survey and calculate area within reasonable accuracy limits, furnishing copy of same to the client and for the public record.
4. To do the work in such manner that a resurvey may be made at an indefinite future time at reasonable cost.
5. To use methods and instruments that will accomplish these objectives at reasonable cost to the client.

A good compass is capable of achieving those objectives for surveys of lower value land.

This author has found *little published data comparing the compass and transit* when used on the same traverse or line. *He has conducted* a number of such studies, and reports on them in this chapter. Except as noted, the compass surveys were made with a 4″ needle open-sight slit-and-hair compass without vernier, graduated in degrees, a tripod, 2-chain steel tape and clinometer. The tape was usually held at knee or crotch height—the rear end being held by the compassman. "Transit survey" means plate azimuths with distances by stated methods. All surveys were made in rough wooded mountain terrain unless stated otherwise.

In measuring long traverses, with many stations, the compass has an advantage over the transit. The law of error propogation for a compass traverse is more favorable than for a transit traverse. The following example is quoted from *Wild Heerbrugg* Leaflet G 1202e 111.66 which describes Wild's TO Compass Theodolite:

> "The accumulation of errors in a compass traverse gives results more favourable than in an ordinary theodolite traverse, due to the independently observed magnetic azimuths. Assuming the same accuracy in measuring the distances, a much higher positional accuracy is obtained with an increasing number of setups, despite the fact that each observed magnetic azimuth is of lower accuracy. This may be shown by the following comparison between the accuracy of positions computed from a theodolite traverse and those from a compass traverse, both running straight. The assumed error of Ms equals ± 0.1 ft. in a 160 ft. leg can be obtained easily when observing a vertical staff with the 1:50 stadia hairs. An angular observational error of Ma equals ± 20″ is assumed in the theodolite traverse, with an error of 0.01 ft. in setting up the instrument and targets, already taken into account. The corresponding figure for the compass traverse are Mw equals ± 1.5′ and 0.5′. The table shows clearly that the relation between the mean square error in x (latitude, along the traverse) and in y (departure, perpendicular to the traverse) is more favourable than in a theodolite traverse with an increasing number of set ups."

ACCUMULATIVE ERRORS

Number of Set-ups (each leg is 160 ft.)	*in x* *theodolite and compass traverses (with Ms at 0.1 ft.)*	*in y* *theodolite traverse (Ma is 15")*	*compass traverse Mw is 1.5 ft.)*
10	± 0.3 ft.	± 0.2 ft.	± 0.2 ft.
20	± 0.4 ft.	± 0.6 ft.	± 0.3 ft.
40	± 0.6 ft.	± 1.9 ft.	± 0.5 ft.

Similar results are quoted by S. E. Huey from an earlier WILD HEERBRUG equipment manual, as follows:

Length of traverse (each side 100 meters long)	*Mean closing error of:* *Transit traverse with error of 20 seconds*	*Compass traverse with error of 2 minutes*
500 meters	± 0.07 m	± 0.13 m
1000 meters	± 0.18 m	± 0.19 m
2000 meters	± 0.52 m	± 0.26 m
5000 meters	± 2.00 m	± 0.41 m
10000 meters	± 5.63 m	± 0.58 m

Both examples indicate that the compass accuracy is superior after about 10 setups.

The Tenney Line

In 1966 a 58.91 chain line was surveyed by compass, and the random stakes offset to the true line between corners. About a year later the line was resurveyed by transit and stadia. The compass survey was on one bearing whereas the transit survey was a series of shorter courses within a few feet of the compass line. The transit stations were tied to 17 of the 32 compass stations.

A true line between termini was calculated from transit data. Then by corrdinates the distance was calculated from the true transit line to each compass stake tied to. The taped distances were used in the calculations because they were probably more accurate than the stadia.

The following table shows the distance *in feet* that each of the 17 compass stations was from the true transit line:

Station	Distance	Station	Distance	Station	Distance
0	0	18	0.47 N	40.5	0.97 N
4	0.17 S	23	1.12 N	42	0.42 N
7.5	0.11 N	25	1.12 N	44	0.67 N
9.5	0.06 N	26.5	1.02 N	48	0.27 N
11.5	0.50 N	28.5	0.11 S	52	0.05 S
15.5	0.71 N	36.5	0.62 N	56	0.70 N
				59.25	0

No compass stations were over 1.2 ft. from the true line. The cost *per mile* for the compass survey was \$28.00; for the transit survey \$76.00 per mile. If a *new* transit line had actually been run, the cost would have been \$80 to \$90. The gain or loss of area produced by the transit survey was 0.04 acre. If woodland in this part of West Virginia is worth \$100 per acre, 0.04 represents land worth \$4.00.

It does not appear sound to spend an extra \$48.00 or more per mile to recover (maybe lose) \$4.00 worth of land. \$48.00 at 6% would return an investor \$2.88 per year, whereas 0.04 acre will produce a timber crop on this particular site worth about \$1.00 per year. Even though land and timber values rise, so also will the cost of surveys. The surveyor must decide whether the presence of coal or other higher values requires a survey by higher-accuracy methods.

Compass and Transit Readings from the Same Station

A compass and transit respectively were set at point (A) in a field, oriented on point (B) (N 43-30 W 21 poles distant). The readings to other stations in the field were:

To Point	*Compass*	*Transit*	*Compass differs by*	*Length of line (poles)*	*Error of Divergence*
C	N 15-45 W	N 15-48.3 W	+ 3.3′	119	+ 1.88 ft.
D	S 18-45 E	S 18-45.0 E	0	39	0
E	S 9-50 W	S 9-41.4 W	+ 8.6′	156	+ 6.42 ft.
F	S 23-30 W	S 23-39.8 W	– 9.8′	116	– 5.44 ft.
G	S 38-35 W	S 38-39.2 W	– 4.2′	100	– 2.01 ft.
H	S 71-00 W	S 70.57.2 W	+ 2.8′	89	+ 1.19 ft.
J	S 89-30 W	S 89-40.6 W	–10.6	95	– 4.81 ft.

The average difference between the divergencies is about 5.6′. The maximum error of 1 in about 326 indicates a limitation of a compass when used on long sights. One reason is that the front sighthair tends to obscure a small target.

A similar test was made with a Gurley No. 227 Compass with slit and hole sights on both vanes. An object about 150 poles away was sighted on from a US MAGNETIC station. After adjustment for declination, the transit and compass readings differed by 5′.

Compare this study with the results in *The Tenney Line*. A compass appears to be better suited for extending a straight line with 2-chain shots than for sighting on objects ¼ mile or more away.

Two Compass Traverses Between Same Termini

One November morning a line was run between two known points. The average slope was about 20°. The line was run on a bearing of S 41

E, and at 24.839 chains missed the marker by 1.21 chains. The calculated true line between termini was S 43-47 E 24.87 chains.

The client then decided to make this line into two lines at a slightly different location. The calculated new lines were to be S 38 E 13½ chains then S 50 E to the corner. In the afternoon of the same day a line was run S 38 E 13.368 chains, then S 50 E 11.653 chains, which missed the marker by 2 inches. A straight line connecting the two termini using the second calls calculated S 43-34 E 24.88 chains.

A difference of 13′ in 25 chains should have provided a miss of about 6 ft. The daily variation could cause some of the difference. The error of 1 in 2490 in distance indicates a precision of 0.20 ft. per tape length.

Two Trial Compass Lines

A compass line was run on a bearing of S 89½ W 30.00 chains, at which point the line was to the left of a marker by 0.935 chain. Terrain was too steep and offsets too long to offset the random stakes easily. The calculated bearing of the true line was N 88-43 W.

So a new line was run on a bearing of N 88-30 (because 43 minutes could not be read with sufficient accuracy on the compass which had no vernier. The second random line was 30.03 chains long and ended 0.125 chain to the right of the marker. The recalculated bearing of the true line from the second set of data was N 88-44 W—just one minute different from the first calculation.

The McClure Line

In May 1967 a compass line was run from an oak stump S 6-15 E 3.486 chains, then south 22.916 chains, from which point a 26″ black gum corner was N 50 W 1.60 chains and N 32-45 W 1.83 chains. This random line was too far from the true line to be offset easily. A plat of the random survey at a scale of one inch to 5 chains showed the true line to be S 4-05 W 23.85 chains.

Several months later a new line was run S 4 W from the oak stump. At 23.90 chains this line hit 0.01 chain east of the center of the gum (within 3 minutes of its expected location).

The Elkhorn Line

About 60 years ago a line was surveyed between two corners as N 86-08 E 439¾ poles. It is not known whether this was a compass or a transit survey. Probably it was the latter.

In October 1966 the corner markers were identified, having been well referenced by stone piles and trees. Some of the witness trees were badly decayed stumps. A resurvey was made on a bearing of N 88 E and the trial line was 1.8 ft. to the left of the corner at 439.73 poles. The average slope was 3 to 4 degrees. The variation tables indicated a change of 1-50 and an adjacent line had shown a change of 2-07.

A Florida Traverse

A closed traverse was laid out in a fairly level field in south Florida. Distances were measured with a 300-foot tape with an estimated error of about 0.05 ft. per side.

A transit survey was made with a Wild 16ED transit with tubular compass, followed by surveys with an old Gurley compass. The compass was in good condition, had a 6″ needle, sight vanes about 15″ apart, 4 levelling screws, 2 level vials and tripod; the vernier was set at zero and *not used.*

Stakes with tacks flush with the ground had longer stakes above them so the tacks were in a vertical line and observable above the high grass. Several stakes were intentionally set near sources of magnetic attraction: a high voltage 4-wire electric line was about 45 feet away from two stations and directly overhead at another; another station was 8 ft. from a pile of angle iron.

The transit's tubular compass was adjusted to the Gurley needle and oriented on magnetic north at a station where no attraction seemed to exist. Transit azimuths were rounded off to the nearest minute and converted to bearings in the table below. The transit traverse was run about 11:00 AM; there was an unresolved angular error of closure of 1.1 minute.

The #1 compass survey was made about 1:00 PM the same day; the #2 and #3 compass surveys were made about 11:00 AM on two later days. Bearings were recorded to the nearest 5 minutes, although several might have been estimated closer.

There was no apparent attraction from electricity (polarity alternates 60 times per second). The angle iron attraction was corrected from backsights. Following are the data from the traverses:

Line	*Length*	*Transit*	*#1 Compass*	*#2 Compass*	*#3 Compass*
AB	185.0	N 31-10 E	N 31-15 E	N 31-10 E	N 31-05 E
BC	245.0	N 89-45 E	N 89-50 E	N 89-40 E	N 89-40 E
CD	168.0	S 8-30 E	S 8-30 E	S 8-40 E	S 8-35 E
DE	110.3	S 23-38 E	S 23-35 E	S 23-40 E	S 23-45 E
EF	227.0	S 88-16 W	S 88-15 W	S 88-15 W	S 88-10 W
FA	216.0	N 57-54 W	N 57-55 W	N 58-00 W	N 58-00 W
	1151.3				

Error of closure computed by latitudes and departures are:

		1:7400	1:2100	1:4900	1:36000

The Ramsey Survey

Two points about ¼ mile apart were connected by a compass survey. The average slope was about 5°. There were 12 stations consisting of 5 courses totalling 22 chains. Later it was necessary to make a transit-stadia survey between the same two points. The transit survey had 10 stations and was run on a slightly different location. A line which would have closed both traverses was calculated as an omitted measurement, with these results:

Compass-tape:	N 25-00 E 20.55 chains
Transit-stadia:	N 24-56 E 20.55 chains

The 4′ difference includes all errors: instrumental, observational, diurnal, etc. Lack of error in distance was probably luck.

The See Survey

A compass traverse was run S 64-30 W from a 32″ white oak, part in open woods and part in a field. At 9.12 chains an N/C was set in the edge of a highway. About two months later a compass traverse was run some distance northerly. The two traverses were then connected by a transit survey to avoid magnetic attraction en route.

The transit was oriented on the northerly compass traverse, and two courses totalling 31.3 chains were run along the highway to the N/C. When the transit was turned to the approximate middle of the large white oak, the azimuth was 64-30. A reading within 5′ of this would have been acceptable.

A Huber Survey

For a 1406-acre survey see chapters on Plats, Descriptions and Other Records.

A Closed Traverse Test

A 5-sided traverse was laid out in a flat field. A Warren-Knight Compass with plumb bob survey was made. Bearings and distances were:

N 57-22 W 100.00
N 4-42 W 100.00
N 57-02 E 100.00
S 27-33 E 100.00
S 17-34 W 125.20

Error of closure was 1 in 16,500. Precisional errors could be these: bob not over tack 1′; pivot not over tack 1′ or 2′; sighting error 1′ or 2′; index error of compass 1′?; reading vernier ½′; and needle reading error 1′. All except index error could be compensating.

A Wagon-wheel Test

Bruce Sumner tested Warren-Knight Compass #17842 as follows: 8 lath 1½″ wide were set at about 45° intervals 200 feet away from the center of a rough circle. The angles between the lath by compass and transit check found the greatest difference to be 4½′ and the average deviation to be 1.6′.

Summary

The accuracies obtained in some of these surveys would be almost impossible to attain in a survey in the woods and sloping terrain. There would be adverse factors tending to reduce accuracy. The value of the **Sipe-Sumner Vernier Compass** lies in its freedom from significant index errors, its ease of adjustment, the simplicity of its use, and its ability to help make surveys within reasonable limits of accuracy at 1/3 the cost of transit surveys.

The final test of a good rural survey is not gauged exclusively by the error of closure of the traverse, but rather by establishing good markers at locations in accord with the weight of controlling evidence and with ties to good reference markers, connecting corners by lines of reasonable or specified straightness and visibly marked, all at appropriate and reasonable cost to the client.

The studies in this chapter should enable a surveyor to determine whether a compass or a transit is the better instrument for the job at hand. We should not allow the geodimeter, the tellurometer, the laser and other new means of measurement blind us to the need for surveys of lower value land that owners can afford to pay for. But to obtain the results described above requires training, suitable equipment, good helpers, ability to master the subject, and experience. The beginner—and even the average surveyor—has much work ahead of him to become *proficient.* All can test themselves by performing experiments and analyzing their surveys.

Chapter 14

TRUE OR MAGNETIC: WHICH TO USE?

Surveyors frequently argue over the merits of running a traverse by true or magnetic bearings. The premise is often accepted that true bearings will produce a more accurate survey. Certain phases of this argument have been discussed elsewhere in this book. The purpose of this Chapter is to explore other aspects of the question.

Let us begin with these assumptions:

1. That a survey in West Virginia was made 50 years ago by magnetic bearings, whether by transit or compass is not known.
2. The surveyor is now required to resurvey the property.
3. He has a transit with needle on which declination cannot be set off, and
4. He has an open-sight compass with a device to set off declination.
5. The first two lines of the traverse are N 60 E and S 24-30 E.
6. The variation since the former survey is 2-05.
7. A transit sun-shot indicates the declination is 5-10 west.
8. In all the following examples, the surveyor hits his objective "on the nose", without meeting obstacles that affect our philosophy (utopia no less).
9. The surveyor has four choices of instruments and methods: A-B-C-D.

A. He selects a transit and decides to run with his plate azimuths. He sets the plates at zero, orients on true north, loosens the top plate. To find the right direction for the first line, he figures thus: N 60 E in the old deed is now N 62-05 E magnetic, or N 56-55 E true. So he turns the scope clockwise to a plate azimuth of 56-55 and runs the line. On the plat he shows two north arrows: the true north arrow is straight up and the magnetic arrow is 5-10 to the left with 5-10 west written between the arrows. On the plat the first line is written as N 56-55 E.

B. With the transit another method can be used. The plates are clamped at 5-10 in the correct direction, true north is sighted on, the top plate is loosened, the telescope and circle turned to 62-05 and the line run. The line on the plat is shown as N 62-05 E. The magnetic arrow this time is straight up and the true arrow is 5-10 to the right with 5-10 showing between.

Now suppose after using method (B), he desires to show on the plat the bearings by true meridian, 5-10 is subtracted (in NE and SW quadrants) or added (in NW and SE quadrants); the true arrow then is shown straight up and the magnetic 5-10 to the left.

In both methods (A) and (B) the actual running is conducted in exactly the same way. Which method is easier?

Both methods are equally accurate. Method (B) appears to have a slight advantage for the man accustomed to think in terms of magnetic bearings and variation. For the man thoroughly familiar with true bearing calculations, method (A) could be just as easy.

Now let us consider the second line: S 24-30 E. For method (A) the surveyor must use two calculations: the variation is *subtracted*, giving S 22-25 E, and the declination is *added* (both just the reverse of the first line) giving S 27-35 E. So he runs S 27-35 E which is shown on the plat.

In method (B), he uses only one calculation—the 2-05 variation—which is turned clockwise from S 24-30 E. He shows S 22-25 E on the plat.

At the end of the second call, method (B) appears to be a little easier to think through.

Now let us suppose the lines are being surveyed with the compass.

C. In this method, the declination of 5-10 is actually set off (extreme care being used to turning it in the correct direction) and the first line is run on N 56-55 E. On the plat is shown N 56-55 E with true arrow up, etc., as in method (A). Other than setting the declination in the correction direction, there is no problem in this method.

D. In method (D) with compass, the declination is not set off, but the line is run N 62-05 E, which is shown on the plat, with magnetic north straight up as in method (B).

Here again in both (C) and (D) declination is being used properly: both methods seem reasonably simple.

Now using method (C), the second line is run. S 24-30 E would be S 22-25 E. With declination set off, to change to true bearing, the surveyor must remember that he has to add 5-10 (counterclockwise) and run S 27-35 E.

If on the other hand, method (D) is followed and the declination is not set off, the surveyor merely has to turn his compass clockwise the amount of the variation; thus he runs S 22-25 E. In this method the compass is turned clockwise in all quadrants.

In all four methods (A-B-C-D), true north and the declination are being utilized properly. But which method is easier—(A) and (C) or (B) and (D)?

(B) and (D) appear to be easier, because:

1. They are simpler to perform and easier to explain to a beginner.

2. When using compass method (D), tampering with the position of the circle with relation to the sight vanes is *avoided*. In method (C), the

adjustment for declination would need to be made for each parcel surveyed. If Smith's tract is being surveyed one day, Jones' the next, and Brown's the third day, and the fourth day the surveyor is back to Smith's, each is at a different place geographically and therefore each could have a different declination.

One cannot escape the conclusion that it is better to allow, in the field work, only for variation, and bring declination into the picture only when making the plat and description.

If this is sound, it leads logically to the conclusion that declination need not be determined out on the line for each tract surveyed (when weather conditions may be unfavorable if not impossible), but can be determined at a *Magnetic* or *Triangulation Station* in the area, at a time convenient to the weather and the observer. The declination at the *station* is recorded on the plat along with its location. This method leads directly to the system recommended by the United States Coast and Geodetic Survey in its booklet Serial No. 718, and as described in the chapter on Magnetic Variation.

With a compass of good precision, the accuracy that can be attained should be satisfactory for the needs of lower value land to be surveyed.

One Final Note: The explanation of the methods described above is not easy to make clear or understand. These explanations can be avoided if declination is utilized as in methods (B) and (D). In fact the transit can be used to run a line "by the needle" just as in method (D), the telescope replacing the open sights.

Public agencies and private companies may of course specify the methods which their surveyors use, within legal limits. For example, the U. S. Public Land Survey system is by law tied to the true meridian. However, since most of the Public Land Surveys up until 1890 to 1894 were executed with the magnetic compass, knowledge of secular and daily variation is indispensible. The logical instrument to use in retracing these old surveys is a good vernier compass, or a transit with a good compass, running "by the needle".

Chapter 15

MAGNETIC VARIATION

Most land surveyors know that there are three kinds of magnetic disturbance that affect the compass needle. They are commonly known as secular, daily and irregular variation. Not well understood however, is how they affect the needle in amount and direction. If the best results are to be obtained with compass needle surveys, the surveyor should know how to adjust or compensate for them. "Variation" should not be confused with "declination" of the needle; the latter will be discussed in another chapter.

Secular Variation

Secular variation is the change in direction of the needle during the passage of long periods of time. There are good records of how much change has occurred in the past. But why the changes occur, and how much will occur, and in what direction, is one of the greatest mysteries of our time. The change in recent years has been from zero to 10 minutes per year. For shorter periods such as from month to month, the change is too small to be considered by the surveyor.

The U. S. Government and other countries operate magnetic observatories to record changes in magnetic conditions and publish data about them. Write or phone Geomagnetism Unit, Branch of Regional Geophysics, U. S. Geological Survey, Boulder, CO. 80302. The magnetic center near Fredericksburg, VA. can provide information for that area; phone 703-373-7601.

The government previously issued Publication 40-2 entitled *United States Magnetic Tables.* The last one issued was for 1960. This publication showed the secular change in the United States from about 1750 to 1960, for every two degrees of latitude and longitude. It also contained other valuable information such as how to interpolate for other locations, and data on daily and irregular disturbances. A new publication is being prepared.

Surveyors have used a rule of thumb of 1° of secular variation for each 20 years. This is too approximate and should be discarded in favor of use of official government tables. Every surveyor should develop his own table of secular variation for the area in which he surveys. A table for this part of West Virginia is shown here. It is adapted from preliminary government data.

Latitude	*38*	*38*	*39*	*39*	*Latitude*	*38*	*38*	*39*	*39*
Longitude	*79*	*80*	*79*	*80*	*Longitude*	*79*	*80*	*79*	*80*
1800	7-04	6-43	7-04	6-45	1890	3-18	3-07	3-15	3-05
1810	7-09	6-48	7-08	6-51	1900	2-42	2-31	2-38	2-29
1820	7-01	6-43	7-01	6-46	1910	2-06	2-08	2-02	1-56
1830	6-45	6-29	6-45	6-21	1920	1-38	1-33	1-32	1-28
1840	6-20	6-06	6-19	6-08	1930	1-04	1-02	0-56	0-55
1850	5-49	5-37	5-48	5-38	1940	0-59	0-58	0-50	0-51
1860	5-14	5-03	5-13	5-03	1950	1-01	1-02	0-55	0-56
1870	4-35	4-25	4-33	4-25	1960	0-43	0-44	0-39	0-41
1880	3-55	3-45	3-52	3-44	1970	0-00	0-00	0-00	0-00
Annual Change in 1970 from Isogonic Chart						4.0W	4.0W	3.5W	3.5W

To bring values of variation up to date, the U. S. Government publishes an Isogonic Chart on the United States every five years. The latest is for 1970.0 (as of January 1). A 1970 chart is shown on the next page.

Interpolation is needed to adjust variations to the latitude and longitude of the survey. The following chart shows how to interpolate from the previous table to obtain variation for an 1890 line to be rerun in 1974 at latitude 38-40 and longitude 79-20.

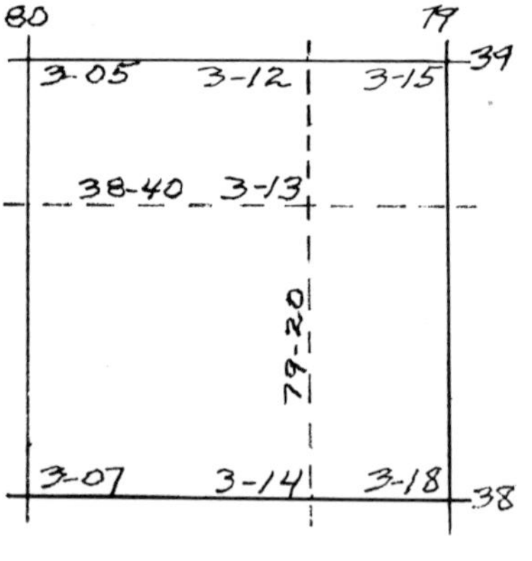

The variations from 1890 to 1970 are shown on the chart. By interpolation, the variation at long. 79-20 and lat. 39 would be 1/3 of 3-15 minus 3-05 or 1/3 of 10 = 3.3. Subtract 3 from 3-15 and get 3-12. At lat. 38 by interpolation, it would be 1/3 of 3-18 minus 3-07 or 1/3 of 11 = 3.7. Subtracting 4 from 3-18 = 3-14. Then interpolating between 3-12 and 3-14, we have 3-13 as the variation from 1890 to 1970 at place of survey. To bring this up to 1974: the annual change in 1970 is 3.5′ per year at lat. 39 and 4.0′ per year at lat. 38. Interpolating, the change is about 3.7′ per year. Then 3.7 x 4 years = 14.8′ or 15′ which added to 3-13 = 3-28. Thus a bearing of S 21 W in 1890 would be run S 24-28 W in 1974.

A wrong understanding of secular change is a cause of poor error of closure when deed calls are plotted. Through the years, as land has been subdivided and transferred, landowners and attorneys have arbitrarily selected calls from deeds (and surveys) made many years apart, and

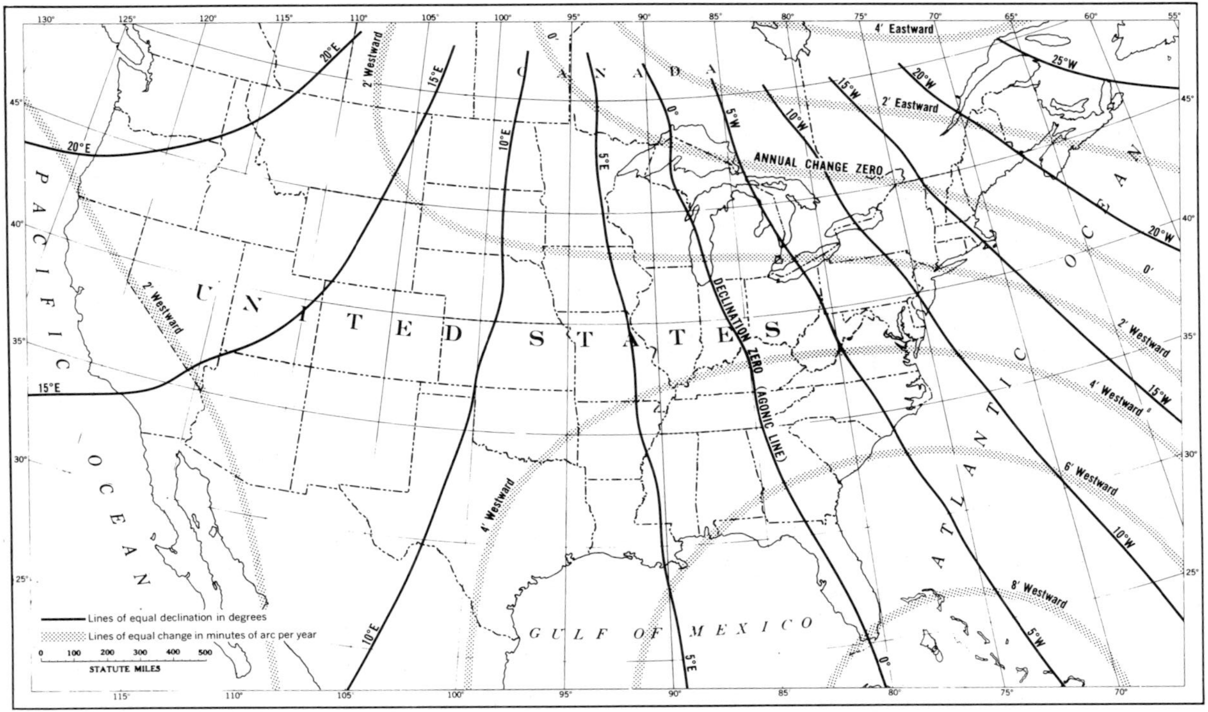

MAGNETIC DECLINATION FOR 1970 (derived from C&GS Chart 3077)

combined them into a new description, not realizing that the needle had changed direction by perhaps 5° or more.

Failure to understand the change also has given rise to arguments over line location. For example, a line may have an old bearing of N 60 E, but if the bearing has changed 5° since that N 60 E call was surveyed, a new survey at N 60 E will certainly encroach on one of the disputing parties. So a resurveyor of a former call must adjust the old bearing for secular variation. In West Virginia the north end of the needle has been moving westerly for many years, so all old bearings would be adjusted clockwise. For example, a S 40 E line in 1850 at lat. 38 and long. 80 would be run at S 34-07 E in 1974. This is arrived at by subtracting the change of 5-53 from the 1850 bearing. The change is 5-37 plus 4 x 4 or 5-53.

If the needle has an eastward movement, the new bearing is counterclockwise from the former one. But it must be understood that in any event, (1) the location of the line *on the ground* has not changed, and (2) the correct location can seldom be decided by a calculation of the variation alone. The government figures are accurate generally to less than 20 minutes, although in later years the accuracy is much closer.

Planning for the Future

The surveyor of today should provide for future secular variation calculations and resurveys of his lines by doing these two things (described in Gillespie's *Surveying* p. 184, and in the U. S. Government Serial No. 718, *Magnetic Surveys* paragraphs 113, 114):

1. Take the compass to a proved true north line in the area several times when the needle is at its mean for the day and average the declination.
2. Record the declination, where and what month and year it was taken, in the notes and on the plat.

Years later a resurvey can be made by:

1. Setting the compass on the same point or true line when the daily variation is at its mean for the day, and reading the declination.
2. Calculating the variation that has occurred and rerunning a trial line on the new bearing.

This system cannot be used to prove the location of a lost or obliterated corner, but will put the crew in close proximity to where it should be sought.

Daily Variation

The compass needle changes its direction not only from year to year, but during the day. Daily variation can be ignored for many compass surveys. But if the straightest line is to be run with compass and the best error of closure to be obtained, daily variation should be allowed for.

The U. S. Government Publication 40-2 and Serial No. 718 show the average daily needle variations for several locations in the United States based on the 10 most magnetically "quiet" days each month. The author has found that the following figures are satisfactory for West Virginia. Probably they are reasonably good for a larger area. The author found them similar to the change at Boca Raton, Florida. The Cheltenham table is shown here.

Standard Time	*Jan. Feb. Nov. Dec.*	*Mar. April Sept. Oct.*	*May June July Aug.*
7AM	1.1 E	3.4 E	5.5 E
8AM	2.3 E	4.7 E	6.0 E
9AM	3.1 E	4.1 E	4.9 E
10AM	2.7 E	2.6 E	1.8 E
11AM	0.7 E	0.6 W	1.8 W
noon	1.6 W	3.3 W	4.5 W
1PM	3.0 W	4.6 W	5.5 W
2PM	3.2 W	4.8 W	5.5 W
3PM	2.7 W	4.0 W	4.4 W
4PM	1.6 W	2.4 W	2.7 W
5PM	0.8 W	1.2 W	1.2 W
6PM	0.2 W	0.7 W	0.3 W

The data formerly recorded at Cheltenham are now collected at the Fredericksburg Geomagnetic Center mentioned before. They can furnish data on daily variation in their area. For other parts of the United States, refer to the Boulder office.

To allow for daily variation, the surveyor should make a chart of the curve for each of the three seasonal groups of data. One such chart of the Cheltenham figures is shown here.

It is obvious that a northerly line run by the needle in the morning would be easterly (cw) of a line that was run just before noon. If continued into the afternoon, the line would pass west of the "mean" line, and if continued till 5 or 6 PM, would return to near the mean line position.

How to Allow for Daily Variation.

1. The first way is to calculate the amount of divergence which the random line would be right or left of where it should be if the needle had been at its mean for the day. Then to allow for this when setting up the compass at each station. The divergence at any station depends not only on the time of day but the distance run.

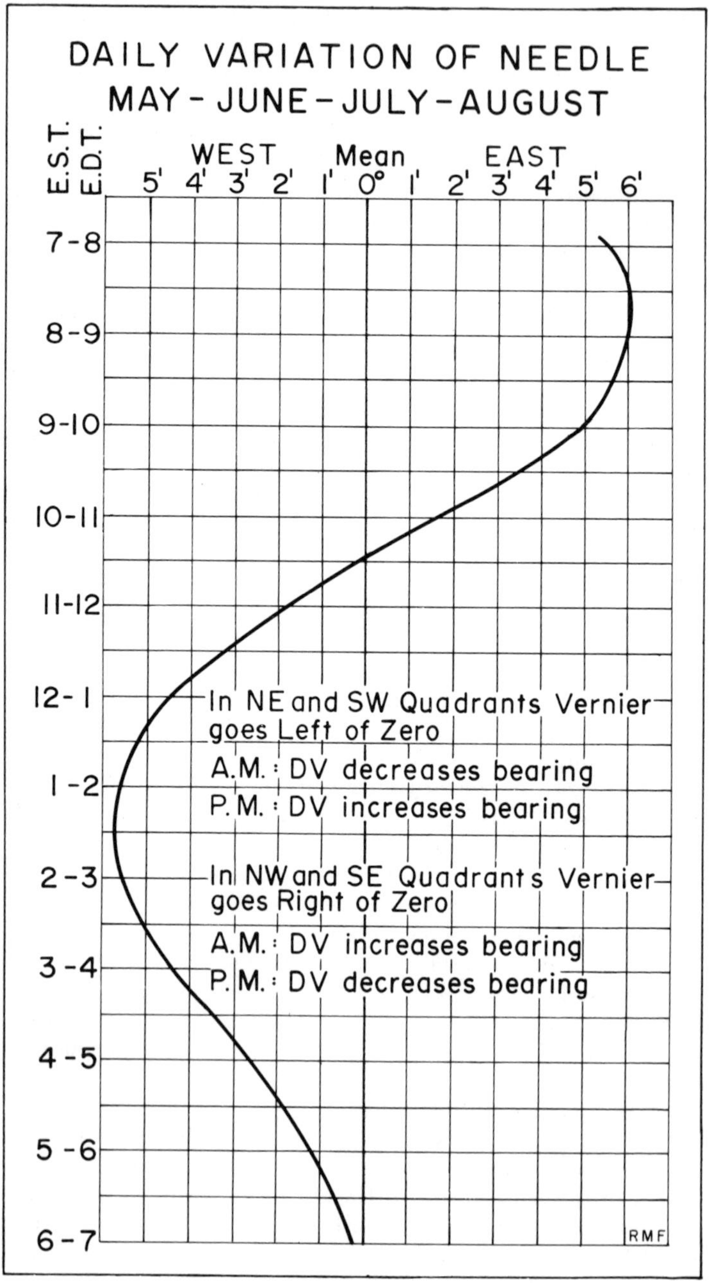
DAILY VARIATION OF NEEDLE
MAY - JUNE - JULY - AUGUST
E.S.T.
E.D.T.
WEST
Mean
EAST
5' 4' 3' 2' 1' 0° 1' 2' 3' 4' 5' 6'
7-8
8-9
9-10
10-11
11-12
12-1
1-2
2-3
3-4
4-5
5-6
6-7
In NE and SW Quadrants Vernier goes Left of Zero
A.M. : DV decreases bearing
P.M. : DV increases bearing
In NW and SE Quadrants Vernier goes Right of Zero
A.M. : DV increases bearing
P.M. : DV decreases bearing
RMF

Assume that a straight line is to be run magnetic north (for ease of explanation) in *May* beginning at 8 AM EST, with ½ hour for lunch, and ending in the afternoon. Assume that about 10 chains per hour will be run. The divergence is equal to the distance run times the tangent of the angle of deviation of the needle from its daily mean. If the deviation is calculated at hourly intervals, the deviation for any hour should be about the average of the deviation at the beginning of the hour and at the end of the hour. The following table (from Serial 718 values) shows the position of the surveyor at intervals during the day, with deviations listed to the nearest ½ minute:

For Hour Beginning EST	*Average Needle Deviation From Daily Mean*	*Which in 10 Chains is:*	*And Puts the Surveyor at*
8 AM	5.5 E	1.06 Ft. E	1.06 ft. E (at 9 AM)
9 AM	3.5 E	0.66 ft. E	1.72 ft. E
10 AM	0.0	0.00	1.72 ft. E
11 AM	3.0 W	0.57 ft. W	1.15 ft. E
12:30 PM	5.0 W	0.96 ft. W	0.19 ft. E
1:30 PM	5.0 W	0.96 ft. W	0.77 ft. W
2:30 PM	4.0 W	0.76 ft. W	1.53 ft. W
3:30 PM	3.0 W	0.57 ft. W	2.10 ft. W
4:30 PM	1.5 W	0.29 ft. W	2.39 ft. W

If these divergences are excessive for the job at hand, the surveyor should allow for them. If the random stakes are offset a proportionate share of the total divergence in an hour's running (10 chains), each 2-chain stake between 8 and 9 AM would be offset 0.21 ft. If the divergencies were included in the calculation for offsets to the true line (as an alternate method), one offset would do both jobs.

2. Method (1) above is too cumbersome. There is a simpler method if the compass is equipped with a one-minute vernier, namely: adjust the vernier each hour to compensate for the deviation expected. S. E. Huey touches on this method in his article of p. 293 of the *ACSM Journal "Surveying and Mapping"* for July-September 1952, Vol. XII, No. 3.

Example A.

Assume for example, a line is to be run N 73-28 W beginning at 10 AM EDT in August. The table shows the needle is 4.9′ east of normal at that hour, which would make the line run, if the bearing was unadjusted, actually be N 73-23 W. To run the proper 73-28 line, at 10 AM EDT the line should be started N 73-33 W. At 10:30 the course should be changed to 73-32; at 11:00 to 73-30; at 11:30 to 73-28; at 12 noon to 73-26, and so on. After lunch the bearing is adjusted again. If the line is started at 11 AM, the bearing is N 73-24 W. In other words the variation is selected from the curve at whatever time exists. In the NE and SW quadrants, the change is added to the bearing run as the needle shifts ccw during most of the day. For the SE quadrant the method is the same as for the N 73-28 W line described above.

Example B.

April 18, 1972 at 1:00 PM EST. Ran line N 6-30 E; no daily variation allowed for. Mean daily variation about 4′ west. Mean line bearing is N 6-26 E.

November 11, 1972 at 10 AM EST. Set up same compass on same line; bearing read N 6-22 E; no daily variation allowed for. Mean daily variation 3′ east. Mean line bearing is N 6-25 E.

Example C.

October 3, 1972 at 11:25 AM EDT. Set compass at point X. Line XY read S 63-21 W; no daily variation allowed for. Mean daily variation is 1½ east. Mean line bearing XY is S 63-22½ W. Then read line XZ as S 48-36 E. 48-36 minus 1½ = S 48-34½ E. Interior angle YXZ is 111-57.

November 13, 1972 at 2:40 PM EST. Repeated the above observations. XY read S 63-20 W which minus daily variation of 3′ = S 63-17 W. XZ read S 48-38 E which plus daily variation of 3′ = S 48-41 E. Interior angle YXZ is 111-58.

If Daily Variation is Not Allowed For.

Assume that a clockwise resurvey of a 50-year-old survey is to be made of a 40-acre rural woodland parcel, with four sides each about 20 chains long, as shown diagrammatically here. The surveyor can do the survey in one day. Beginning at A at 8 AM, he will reach corner B at 10 AM. If he fails to allow for daily variation, he will be 1.72 ft. to the right of his objective (see second table). If the marker is present, he ties to it and can later offset his stakes, allowing the proper proportion of the 1.72 ft. "miss" (plus or minus any other "miss"); failure to adjust the vernier is not significant. Beginning the second line at found marker B at 10 AM, the surveyor can expect to be 0.57 ft. left of corner C. If this and other markers are found, he proceeds as with the first line, and failure to adjust for variation is not significant.

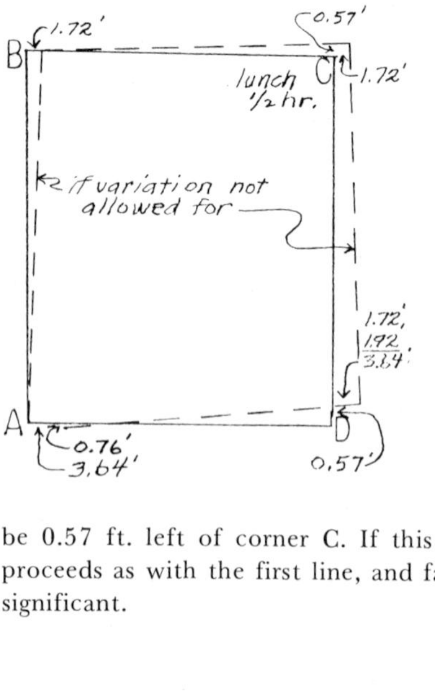

If *only* the beginning marker is found, the surveyor who does not allow for daily variation could expect to find himself at the respective positions shown in the diagram, namely: at corner B—1.72 ft. to the right; corner C—1.72 ft. too far and 0.57 ft. to the left; at corner D—1.72 ft. plus 1.92 ft.=3.64 ft. to the left and 0.57 ft. short. The closing line would fall 3.64 ft. short and 1.33 minus 0.57 = 0.76 ft. to left of the proper point.

Geomagnetic Storms.

There will be days when the daily variation does not follow the normal curve; these are called "stormy" days. The average divergence of 60 "stormiest" days in 1962 was 0.2′ different from the average of 120 "quiet" days, but 1962 may have been a "quiet" year. Also an average for a stormy day could represent an abnormal divergence in one direction in the morning and in another direction in the afternoon, and the surveyor may be making his measurements during a peak or depression which occurs during disturbed periods. But the abnormality of most days will not likely be great enough to prevent the use of the "quiet-day" curve. The surveyor does need to watch for those days when a geomagnetic storm of unusual intensity is occurring or predicted. How is this done?

The Boulder office publishes each week a forecast of geomagnetic conditions and a record of past conditions. This publication is sent air mail on request to interested parties. A summary of the week just past is given, then a forecast for the coming week and for the next 27 days. The geomagnetic field forecast is either quiet, unsettled or active. Geomagnetic storms are minor or major. When the so-called A-index is greater than 30 and less than 50, a minor storm is in progress; if greater than 50 a major storm.

The surveyor has an additional source of more up-to-date data. He can tune a shortwave radio to 2.5, 5, 10, 15, 20 or 25 MHz and receive information at 18 minutes after every hour.* The geomagnetic conditions for the day before, and the forecast for the current day are given. This information comes from the National Bureau of Standards Radio Station WWV at Fort Collins, Colorado. Reception varies depending on weather and which channel is tuned in. If reception is bad, the same data can be obtained by dialing direct to 303/499-7111 at about 17 minutes after the hour; the live broadcast is tapped at its source just like phoning for the correct time. The time is given every minute by WWV. (Time signals are also broadcast at 7.34 MHz from Canada.)

A rare geomagnetic storm occurred from August 3 to August 6, 1972. It received wide publicity in the news media. The A-index rose to 139 on August 5 at Fredericksburg. A needle survey made during the storm would likely have shown odd results. The author did not take any observations during this storm. He was alerted to a less severe storm

*WWV also gives current geomagnetic conditions at 14′ after the hour.

on September 13, 1972 and was able to set up a compass and take readings during that day. The A-index at Fredericksburg was 25. The following table shows the daily change in declination of the needle.

HOUR	*AT FREDERICKSBURG*		
EST	*NORMAL*	*ACTUAL*	*AT ELKINS*
10 AM	7-29	7-40	
11 AM	7-33	7-38	5-36
12 noon	7-36	7-35	5-38
12:30 PM	7-37	7-32	5-32
1:00 PM	7-38	7-35	5-37
2:00 PM	7-38	7-40	5-42
3:00 PM	7-37	7-46	5-49
4:00 PM	7-35	7-45	5-48
5:00 PM	7-34	7-37	5-41
6:00 PM	7-33	7-34	5-37

If the Elkins readings are curved and superimposed on a curve of the Fredericksburg actual curve, it will be seen that they match quite well. The divergence from normal indicates that some compass surveys scheduled for that day should have been postponed.

The Fredericksburg Geomagnetic Center, Corbin, VA., will give information on these storms by phone (703/373-7601) upon request. They will also send a "sensitive" magnetogram chart of geomagnetic activity for specific days upon a request made a few days after the date desired. Their charts are sent to Boulder soon after the end of the month. Boulder charges 60¢ each for copies of charts. Microfilm copies of magnetograms for one year cost $5.00 (35 mm. size).

The reader of this chapter up to here may conclude that the compass needle is rather uncertain. This is not generally true. Geomagnetic storms of sufficient intensity to affect most compass surveys greatly are rare. For example, note the Tenney survey in Chapter 18 which was made without any knowledge of geomagnetic conditions. A reasonable conclusion is that the main reason for allowing for daily variation is to survey the *straightest* practicable line between two points; it will also improve the error of closure. If a survey of second or third order accuracy is being retraced, or if third order accuracy is desired, daily variation should be allowed for. The longer the line, the greater the benefit by allowing for daily variation.

A Summary: Allowing for Daily Variation.

1. Make chart of daily variation for the three seasons (for field use).
2. Obtain the Boulder weekly preliminary review and forecast of magnetic activity; keep abreast of conditions.
3. Obtain magnetic activity of "day before" and current day by

tuning in to radio station WWV on shortwave channel. If reception is bad, phone 303/499-7111.

4. If conditions are "go", tune in or phone just before 11:18 PM EST for confirmation of activity next day, or tune in or phone early on the day of survey just before 18 minutes after the hour.

5. If no storm is in progress or forecast, make the survey. Use the relevant mean daily variation curve and adjust the vernier at time of starting the job and at periodic intervals thereafter.

6. If storm exists or is forecast, postpone the survey.

7. If survey is made but a significant *geomagnetic* storm did occur, obtain from Fredericksburg or Boulder a sensitive magnetogram for the day of survey. Calculate (a) divergence of actual daily variation from the daily variation used, and (b) the random stake offsets to allow for the divergence.

Local Magnetic Disturbance

Natural Disturbances.

The compass needle occasionally varies by a few minutes for places 100 feet apart; such variations are related to local geologic structure. Where iron ore deposits exist, the disturbance may be great and make compass surveys more difficult if not impractical. The author has little experience in such areas.

Artificial Disturbance from Surveyor's Equipment.

Iron and steel will affect the needle. Most such sources are of a "dipole" nature and their effect is only 1/8 as large if the source is twice as far from the instrument. The surveyor should know how far to keep the compass from these sources of disturbance. He should carry as little magnetic material on his person as possible.

Authorities are vague on this subject. Sources of information are: *Davis & Foote* (1940 page 254); *Tracy* (1907 page 543); *Army Technical Manual* TM-5-232, (page 167); *Gillespie* 1920; and *Hodgman* (1903 page 27). Probably the best description of disturbance is in *Magnetic Surveys* (National Geodetic Survey Serial No. 718 page 8). It points out clearly that alternating electric current does *not* affect the needle.

Small Tools.

The author conducted a number of tests of small items, all with the Warren-Knight Compass No. 130. The compass was rotated so the needle pointed northerly on a 10° graduation mark and the vernier was set at zero. A 4X magnifier was used to detect needle movement. The observer stood at the south end of the needle in a normal position. If the north end of the needle moved away from the 10 degree mark during a test, the vernier was used to bring them together and the divergence was read on the vernier.

Spectacles.

Some glasses contain no magnetic material and others will move a compass needle 5-10 minutes. Steel hinge pins are a source even though frames are plastic. One pair of the author's moved the needle 7 or 8 minutes ccw (counterclockwise) when 10 inches from the north end of the needle. Nonmagnetic glasses must be worn.

Magnifier.

Some magnifiers have ferrous pivots or eyelets and must not be used. A 2-inch diameter magnifier with a 2-inch handle will keep the hand from bumping the compass.

Vest.

A red Filson cruiser vest Model 10 has metal snap fasteners on the front. These are slightly magnetic but are compensating when the observer is south of the south end of the needle and when the vest is open. No visible disturbance was detected from the fasteners at a normal 12″ from the needle.

First Aid Kit. (2½ x 4 x 7/8″

When carried in the back vest compartment 2 feet from the needle, no attraction was noted. A snakebite kit may be nearly all plastic.

Cap.

Some bump caps have metal fasteners. If the cap is on the head above the needle, no disturbance may occur.

Field Book.

Looseleaf books with metal binding cores must be kept 12″ to 14″ from the needle. If carried in a side front vest compartment it sometimes comes too close to the needle.

Belt Buckle.

The normal lightweight buckle will not move the needle when in a normal position about 12″ away from and below the needle. However, the buckle is sometimes closer than that when running down hill. It is best to use a brass or non-magnetic buckle.

Knife.

Knives vary in their magnetic effect. Disturbance is often noted when they are two feet away. Do not carry a knife.

Pruner.

A Stanley PR 19 shrub pruner is mostly plastic and has no adverse effect until it is within 5″ or 6″. It may safely be carried in a hip pocket. A True Temper Rocket A-35 will affect the needle at 15″.

Watch.

Watches vary depending on what they are made of. A Timex with Speidel metal band disturbed the needle at 12″.

Car Keys.

They also vary in magnetic effect. Each user must test.

Pencil.

A Scripto with plastic cover will move the needle when within 7″. Use a pencil without metal.

Pen.

A 5″ ballpoint pen with plastic cover moved the needle at 5″. No disturbance noted when 10″ away from needle.

Clinometer and Abney Level.

A Suunto clinometer and Warren-Knight Abney Level had no disturbance.

Hand Axe.

A 28 oz. solid metal handle axe with rubber grip caused disturbance when 4 feet from the needle.

Steel Tape.

A ¼″ tape when stretched parallel to needle, moved it when within 20″; none at 30″ below the compass.

Comment.

The above tests show that some metal objects can be carried by the compassman and used normally without visible disturbance to the needle, *provided* he keeps a normal distance from the compass. Some of them are compensating. It is wise to use field books without metal cores, or carry them in the back vest compartment. The maximum items that can safely be carried are cap, vest, pruner and first aid kit and these must be kept at safe distances. Others like watch and car keys might be carried below the belt line. A brass belt buckle should be used. The danger with any of these items is that unusual terrain may place them too close and the needle will be affected. The compassman must constantly be alert against getting within danger range. The fewer magnetic items he carries, the less is his worry. Fortunately some residual effects totalling not over a minute or two are constant throughout the survey. Occasional slipups are insignificant and non-cumulative.

An Example of Variation and Disturbance

A surveyor calculated the bearing of an omitted measurement of an open-end traverse made the previous year as N 73-28 W. He ran the calculated 5050-foot line with the same Warren-Knight opensight compass and missed the marker 19.3 feet to the left. This is equivalent to a miss of 13 minutes. The surveyor had adjusted the vernier every half hour to allow for the mean daily variation.

Seeking the reason for the 19.3-foot miss, the surveyor found that:

1. He failed to allow for a 4′ secular variation for the one-year period, and
2. His spectacles moved the needle 5-10′ ccw when in the normal observing position.

The surveyor then offset the random stakes and was able to check a

1200 ft. portion of the line through a field. He found the bearing at 3:30 PM EDT to be N 73-27 W. Allowing 4′ for daily variation from the mean at that hour, the adjusted observed bearing was N 73-23 W.

Assuming that the needle was moved an average of 7′ ccw on the *random* line, instead of running N 73-28 W, the surveyor actually ran N 73-28 W plus 7′ or N 73-35 W. Then N 73-35 W minus N 73-23 W = 12 minutes. This error would be accounted for by the 4′ secular variation for one year plus the disturbance from eyeglasses of 7 or 8′.

If the surveyor had allowed for these two factors, he could have expected to come within 1-3 feet of hitting the marker. The offsets would then have been easier to make.

This line was 70% wooded with slopes up to 33°. The random line was run by 3 men in 4½ hours. The random line was offset, marked (including 6 pipes set) and painted by 3 men in 4½ hours. The times are "on-the-job" hours.

Artificial Disturbance Met on Survey

Fences, railroad tracks, culverts, pipes, automobiles and other objects are frequently encountered on land surveys. Tests have been made to see how far to stay from them. The best way found to test fences was to set a line of stakes at right angle to them, keeping stakes in line. The compass is set at a stake at one end and the bearing of the line of stakes read. Then move to the next stake and repeat the reading.

Wire Fences.

A compass was set up NE of a rusted fence with 6 woven and 3 barbed wires. The fence ran about S 62 E. The needle was disturbed cw within 15 feet of the fence. A needle SE of a rusted fence of 8 woven and one barbed wire which ran N 36½ E. A ccw disturbance existed within 20 feet of the fence.

Several tests were made with a Warren-Knight compass by Prof. Darrell R. Dean Jr. of Glenville State College. With compass SE of a fairly new 30″ woven wire fence with one barbed wire above and one below, and with metal posts, there was disturbance up to 16 feet away (the maximum distance checked). With compass NE of a similar fence running N 74 W, the needle showed almost no disturbance within 4 or 5 feet of the wires. This relative lack of disturbance may have been due to the needle being almost at right angles to the fence and the polarity of the wire.

Railroad Tracks.

The compass was tested NW of a 90 lb. railroad track running N 39 E. The distances and deflections of the needle were:

Distance	*Deflection*	*Distance*	*Deflection*	
10 ft.	600 min.	50 ft.	14 min.	The needle was moved ccw.
20 ft.	194 min.	60 ft.	4 min.	
30 ft.	72 min.	70 ft.	1 min.	
40 ft.	34 min.	75 ft.	0 min.	

A similar test was made SE of the same railroad on a different day with these results:

10 ft.	197 min.	50 ft.	6 min.	
20 ft.	103 min.	60 ft.	2 min.	The needle
30 ft.	42 min.	70 ft.	1 min.	was moved ccw.
40 ft.	20 min.	80 ft.	0 min.	

The intensity was only half as strong SE of the track.

New Wire vs Rusted Wire.

Two 10″ pieces of No. 9 galvanized wire (one new and one rusted) were placed parallel to the needle and 6″ away to the east, then reversed end for end. They were then placed at right angle to the north end of the needle and 6″ away, then reversed end for end. The same test was done with a piece of new No. 12 wire and rusted No. 12 wire each 10″ long. The needle moved as shown in the table below:

	6″ away & parallel		*6″ away at a right angle*	
Wire	*Position A*	*Position B*	*Position A*	*Position B*
No. 9 new	14′ ccw	6′ ccw	1°-15′ ccw	1°-10′ cw
No. 9 rusty	8′ cw	5′ cw	12′ ccw	7′ cw
No. 12 new	15′ ccw	12′ ccw	12′ cw	6′ cw
No. 12 rusty	10′ ccw	5′ ccw	9′ cw	0

In this test, rusted wire made less disturbance than new wire of the same size.

On the other hand, the following test showed different results. Two 24″ pieces of 2-strand barbed wire about 14 gauge, one new and one rusted, were tested in the same way with these results:

	6″ away and parallel		*6″ away at a right angle*	
Wire	*Position A*	*Position B*	*Position A*	*Position B*
No. 14 new	35′ ccw	21′ ccw	1°-25′ ccw	1°-00′ cw
No. 14 rusty	2°-00′ ccw	2°-00′ ccw	1°-20′ cw	12′ cw

The angle of the wire makes a difference.

Automobiles.

A 15-foot long Plymouth Duster was parked in a field headed NE. A line of stakes was set almost north beginning 82 feet south of the car's center, passing 5 feet west of the left rear fender, and continuing for 50 feet north. There was no needle disturbance until within 60 feet on the south and 50 feet on the north. A 16½-foot long Plymouth Satellite parked headed SW gave no disturbance to a compass 50 feet north of the right rear fender. The needle when nearer the car was moved ccw.

In contrast, a compass sitting 25 feet east of a lane of vehicles which moved rapidly north, was undisturbed. The needle did not have time to react.

Electric Line.

A line was run under and at right angle to a 12KV AC line with wood poles which ran N 39 W. There were 4 wires in a horizontal plane about 25 feet overhead, and 4 wires in a vertical plane 35-50 feet above. A similar test was made under a line running N 43 E. There was no disturbance to the needle. *Direct* current will disturb the needle.

Gas Pipe Line.

Three pipe lines were 30 feet apart and 5-6 feet below ground. Bearing of the lines was S 76 W. The north pipe was 36″ in diameter and the two south pipes 26″. Needle disturbance disappeared 45 feet south of the south pipe and 110 feet north of the north pipe. Some companies use large amounts of DC electricity to prevent corrosion.

Metal Culverts.

Metal culverts are made in many sizes and the author has not tested a wide variety of them. Their disturbance to a compass needle depends not only on their size, but on their position and distance from the needle. In one test a line of stakes was set about east across a 22″ x 24′ 10-year-old corrugated iron culvert near the surface of the ground. The bearing of the line from each stake was as follows:

over the culvert	– N 85-13 E
2½′ east of	– N 85-53 E
5′ east of	– N 85-35 E
10′ east of	– N 85-30 E
15′ east of	– N 85-30 E
40′ east of	– N 85-32 E

Contrast this with a compass 5 feet NE of the NW end of a 24″ culvert; the needle was moved 1° 36′ cw.

Directly over another culvert (20″) near the surface, the needle was moved 40′ ccw; at 8 feet SW of the culvert, it was moved 29′ cw; and at 8 feet NE of the culvert, it was moved 11′ ccw.

On an actual survey job a line was run from an undisturbed point (A) S 85-30 E 2.00 chains to pojnt (B), crossing a 60-foot lowwater bridge which ran SW to NE and contained 15 corrugated iron culverts about 2 feet in diameter. Point (B) was about 30 feet SE of the bridge and about 30 feet south of the NE end of the bridge. A line was to be run from (B) to point (C) northerly. Anticipating disturbance from the culverts, the interior angle ABC was calculated as 69-45. When the compass was set at (B), a backsight to (A) read N 84-40 W, showing a disturbance of 50 minutes ccw. The 69-45 interior angle was turned and

2.00 chains run at N 14-55 W to point (C). Setting up at (C), a backsight to (B) read S 15-45 E which it should have been.

Iron Pipes.

Tests made with a compass at varying heights over a ¾″ x 34″ black iron pipe in the ground indicate there was no disturbance until it was as close as 30″, provided the pipe was vertical, and directly under the pivot. However, if the pipe leaned, the needle was disturbed.

A 12 ft. ¾″ iron pipe was laid in a N-S position on the ground 48″ below the compass, with the middle of the pipe under the compass. It did not disturb the needle. There was some disturbance when the pipe was moved easterly but still parallel to the needle; this disappeared at about 10 feet.

The same pipe was placed in an E-W position under the compass. There was no disturbance, but when moved north or south and still in an E-W position, there was disturbance up to 10 feet.

Comment.

These tests show that the needle of the Warren-Knight Compass No. 130 was disturbed when used closer than these distances to the objects listed:

Rusty wire fence	– 15-20	feet
Newer wire fence	– 20-25	feet
automobile (compact)	– 50	feet
90 lb. railroad tracks	– 75	feet
gas pipe lines	– 110	feet
iron culverts and pipes	– variable depending on size, position and distance.	
electric lines	– AC current–no disturbance.	

When offsets are not possible and the compass must be used closer than these distances, the surveyor must rely on interior or deflection angles, or project his line from backsights. Each surveyor should keep a special file of his own experience with sources of disturbance.

The question arises: what if there is a hidden source of disturbance or an unknown magnetic storm; how can it be detected? The best way is to see if the back stake lines up with the bearing being run. The compass is not rotated 180°. A 5-minute deviation at 2 chains will make a 0.2 foot divergence of the line of sight. This is usually detectable at 132 feet, and should be reason for seeking the cause. Sometimes stakes are not vertical. If there is a hidden source of disturbance it should disappear at the next station. The technical ability of the surveyor will be tested to find the trouble. He may wish to blame the instrument when more likely it is something that he is doing wrong or has overlooked.

Chapter 16

INDEX CORRECTION OF A COMPASS NEEDLE

How much index error exists in the compass being used? *Magnetic Surveys* Serial No. 718 (p. 18) states: "It is advisable to take two or more tests on different days and at different hours (preferably early morning and late afternoon)." This job can be a little complicated. The following steps were performed at a *Magnetic Station* at Ft. Lauderdale, Florida (Plantation) Lat. 26-06.4; Long. 80-12.9.

In March 1968, a transit was set over the station. The plates were clamped at the azimuth of a reference mark and the telescope sighted on it. The upper plate was released and checked on the other reference marks. Then the telescope was turned to 180-00 (true north) and an ½″ x 8″ rod set flush on the ground 35.6 ft. (at a convenient spot) from the station. A pin was set behind the rod; both appeared correct within 0.005 ft.

A No. 227 Gurley Surveyor's Compass (6″ needle) was set over the station with plumb bob, and tripod legs firmly set. With vernier at zero the compass was sighted at the pin and clamped; it was levelled and resighted with tangent screw. The needle read about N 00-20 E. By adjusting the vernier, the zero on the upper plate was brought to the needle, and the vernier read and recorded. This operation was repeated throughout the day.

A key chain was lowered to the north end of the needle so as to rock it slightly in a vertical plane. The rocking was repeated several times for each reading. A 2-power glass was used to observe the needle and vernier. The north end of the needle was always read, with the observer standing at the south end. The total error of setting up and observing appeared to be less than one minute. An observation was taken every hour with the results below.

Time	*March 6*	*March 13*
9:00	0-20	0-22
10:00	0-20.5	0-23
11:00	0-18	0-23.5
12:00	0-21	0-25
1:00	0-23	0-27
2:00	0-25	0-27
3:00	0-26.5	0-27
4:00	0-25.5	0-26
5:00	0-26	0-24.5

These data were sent to the U.S. C.&G.S. for analysis. The index error of the compass was said to be "-3 minutes of arc to west declination." This means the needle points 3′ to the west of where it should point. So if the observed value of the declination at a given point is 5-20 west, the true value is 5-17 west.

Chapter 17

THAT CLINOMETER—HOW WELL DOES IT WORK?

The measurement of slope angles is one part of the technique of land surveying. Slope angles are used to reduce slope distances to horizontal. The open-sight compass surveyor must measure the slope or vertical angle with a clinometer; breaking chain for horizontal taping is seldom desirable for long lines.

Instruments Used

The author has used two types of clinometers—abney levels and Suunto (pronounced Soo-un-to) clinometers. How accurate are they in themselves and how accurately can they be used?

Manufacturers do not specify the inherent accuracy with which their instruments are made. One company states that theirs "can be read" or interpolated to 5 or 10 minutes, particularly in the "near zero" area. The results obtained by the surveyor are a combination of instrumental index error and human error.

An error of 10 minutes for gently sloping land will not result in a significant horizontal error. For example, a 10′ error on a 10° slope for a 100-foot distance will result in a 0.05 ft. horizontal difference; this is an error of 1 in 2000. The error is mostly compensating and is due to unsteadiness of hand, interpolating and inability to sight on an object or point exactly at the same height as the observer's eyes. Therefore in a mile of line with 53 one-hundred-foot measurements, the total error is the single error of 0.05 x the square root of the number of opportunities or 7.28 x 0.05 = 0.364 ft. or an error of 1 in about 14,500.

Even at 35° a 10′ error seems acceptable. For example, a 10′ error (35°10′) gives a horizontal error of 1.24 ft. in 53 compensating observations. 1.24 in 5280 = 1 in 4250, which is well within fourth order work. However, if the error is much greater on the steeper slopes, the work may be unacceptable. To see what results can be obtained, the author made certain tests with several Suuntos and a Warren-Knight abney level under "backyard" conditions, and with a Suunto under field conditions.

Backyard Tests

After experimenting with testing methods, the author settled upon this: a 15.8 ft. rod marked in feet was set reasonably vertical 29 feet from an open garage door which was at ground level. Two 8-ft. rods were laid end to end from the base of the 15.8 ft. rod toward the open door. The rods were equidistant from both sides of the door and provided targets from + 20½ to −20½ degrees. A tree top was sighted for a steeper plus angle.

A transit was set in and near one side of the door. A small wood shelf was screwed to the other door stop at transit height. The shelf could rotate and the Suunto could be set on it. A small piece of wood was taped to the side of the shelf so it could be flopped on top of the shelf for the abney level to sit on (it has a lower observing line). Repeated readings were made on various footmarks on the rod with a transit, two Suuntos and the abney level. Readings were made with one eye and two eyes, in 85° temperature and with instruments just out of the refrigerator.

Results of the tests showed that there was a difference between Suunto values read with one eye and two eyes. The author also noted that when the Suunto scale was "stared at" for a few seconds with two eyes, the reading was likely to fluctuate. The Suunto Company instructions that formerly recommended two eyes now state that some people have "heterophoria"—a condition where the axes of the eyes are not parallel. The best technique for the Suunto is to locate the object with two eyes and get the clinometer into final position, then close the left eye, turn the Suunto slightly to the right and make the final reading with the right eye. It was also found that test results would be adequate if every footmark was not measured, but instead use only a few plus and minus points. The following chart shows the results of five observations on five rod footmarks and a treetop:

	+39-17	=	TRANSIT	=	+20-34	
Suunto 1	*Suunto 2*	*W-K abney*		*Suunto 1*	*Suunto 2*	*W-K abney*
39-05	39-10	39-00		20-35	20-10	20-25
39-15	39-00	39-00		20-40	20-10	20-20
39-10	39-05	39-05		20-45	20-10	20-25
39-15	38-55	39-10		20-35	20-15	20-30
39-15	39-00	39-05		20-40	20-10	20-30
39-12	39-02	39-04	(Average)	20-39.4	20-11	20-26
-5	-15	-13	(Deviation)	+5.4	-23	- 8
	+ 9-37	=	TRANSIT	=	0-00	
9-35	9-25	9-40		+0-10	−0-12	+0-02
9-45	9-30	9-30		+0-05	−0-10	+0-02
9-50	9-30	9-30		+0-13	−0-13	0-00
9-50	9-30	9-40		+0-10	−0-12	+0-01
10-00	9-30	9-40		+0-05	−0-14	+0-02
9-48	9-29-2	9-39.4	(Average)	+0-8.6	−0-12.2	+0-1.4
+11	- 8	+2.4	(Deviation)	+8.6	−0-12.2	+0-1.4

	- 8-07	=	TRANSIT	=	-20-34	
8-00	8-25	8-00		20-35	20-40	20-30
7-55	8-20	8-00		20-40	20-45	20-25
7-55	8-20	7-55		20-35	20-35	20-30
8-00	8-15	8-00		20-30	20-40	20-30
7-50	8-20	7-55		20-30	20-40	20-30
7-56	8-20.6	7-57.8	(Average)	20-34.4	20-32	20-29.2
+11	-13.6	+9.2	(Deviation)	- 0.4	+ 2	+ 4.8

The deviations suggest that the user might make certain allowances when using specific instruments in the field. For example, Suunto #2 and the abney could be raised a little on plus slopes. But first be sure that all the adjustments listed by the abney manufacturer are made.

Two other hand-held clinometers were examined but not tested by the author: (a) Breithaupt's *Necli* is liquid-damped and the image of the object appears within the field of view of the graduations. Its higher cost may not insure a correspondingly better result than the Suunto. Also it does not fit conveniently in a breast pocket. (b) The Freiberger Pendulum clinometer costs less than the Suunto, but the graduations are too close together and the opaque window may not let in enough light.

Field Test

The author has made one good field test comparing a Suunto with vertical angles by transit. The horizontal accuracy of this survey was discussed on page 133 of the former book as "The Tenney Line", and is also described in this book. The field notes for the two Tenney surveys of the same line show these slope angles (but not with the same Suunto as in the "backyard" tests):

Station	*Transit*	*Suunto*	*Devi-ation*	*Station*	*Transit*	*Suunto*	*Devi-ation*
109-107	+ 3-32	+ 3-40	+ 8	80-78	+ 0-48	+ 0-30	+12
107-103	+ 3-36	+ 3-30	- 6	78-76	+ 5-03	+ 5-00	- 3
103- 99	+ 3-42	+ 4-00	+18	76-74	+ 5-06	+ 5-30	+24
99- 95	- 6-22	- 6-30	- 8	69-67	- 7-10	- 7-30	-20
95- 93	-13-54	-14-00	- 6	67-63	+ 2-36	+ 2-30	- 6
93- 92	+ 0- 4	0	+ 4	63-61	+ 0-30	+ 1-00	+30
92- 90	+ 0-26	+ 0-30	+ 4	61-59	+ 2-16	+ 2-00	-16
90- 88	+ 0-54	+ 1-00	+ 6	59-55	+10-43	+11-00	+ 7
88- 84	+ 1-26	+ 1-30	+ 4	55-52	- 2-40	- 3-00	-20
84- 80	+ 0-46	+ 1-00	+14				

Total + deviation = 131 Net deviation = 46
Total - deviation = 85 Per observation = +2.4

In this case the author was resurveying a line that had last been surveyed about 60 years before. The former distance was 439.75 poles. The author's distance was 439.73 poles—miraculously close to the former distance. Unfortunately there were too few steep slopes in this test.

In Summary

It appears that Suunto clinometers and the Warren-Knight abney level are satisfactory for determining slope angles for fourth-order work *when properly tested, adjusted and used.* If the front rod-chainman and compassman each have a clinometer, an average of the two readings should improve accuracy. The objective is to keep the error of observations within 10 minutes, especially on steep slopes. The author has suggested to the Suunto Company that half-degree lines be added to the degree side of the scale.

The abney level tested gave generally more consistent readings than the Suuntos, and might have been easier to read if it had a vernier. An abney requires more time to read in the field and a magnifier may be needed.

Each surveyor must test his clinometers and level periodically. A permanent shelf can be erected and several plus and minus points selected for observation. The same observing techniques, glasses, etc., that are used on the job should be used in the tests. An extra abney and Suunto could be kept on hand.

The Suunto Company advises that (a) a bubble in the clinometer does not affect the reading if it is not in the field of visibility on the hairline; (b) like a watch, the Suunto can be damaged by dropping or by other heavy shock and that if a rattle is heard, it should be checked for accuracy, and (c) clinometers may be sent to the factory at Suunto Oy, SF-02920, VANHAKARTANO, Finland, or to the importer (Forestry Suppliers) for repair.

Finally it must be remembered that in the field, unsteadiness in holding the clinometer and its interpolating errors will likely outweigh the index error of the instrument.

Chapter 18

DETERMINING DECLINATION

"Declination" (as distinguished from "variation" of the needle) is the angle between true north and magnetic north at any given time and place. If all surveys were oriented on true north, declination would not be a problem. Until such time as true bearings are required and in general use, a knowledge of declination is essential. Even when true bearings are known, many surveyors will relate these to magnetic bearings by use of declination.

How to Find the Declination

The compass must be placed at a point from which the true bearing of a mark or line is known.

At Your Own Station.

The surveyor can establish his own true north line or a true azimuth to selected marks. The station selected for the instrument should be as free as possible from magnetic disturbance. Check this by setting several stakes in a line; if the bearing of the line is the same at all stakes, there is likely little or no disturbance, and any of them may be selected. If disturbance is found, additional points must be checked. Finding true north is described in another chapter. Stations of other surveyors or agencies should not be used until checked.

At A Public Marker.

The National Geodetic Survey, which is a component of the National Ocean Survey (formerly the U. S. Coast and Geodetic Survey), is responsible for the establishment and maintenance of this nation's horizontal and vertical control networks which today consist of approximately 140,000 and 430,000 marked control points respectively. NGS publishes the data related to these stations and makes it available to the public at a low cost per sheet. A horizontal control data sheet provides descriptive data regarding the station, reference, and azimuth marks; geographic and plane coordinate positions for the station, elevation, if available; and sufficient information for obtaining geodetic, plane coordinate, or true azimuths from the station to the reference and azimuth marks. Vertical control data sheets will provide the descriptive data and elevations for vertical control points.

These data sheets, or information regarding the existence of such control in your area, may be obtained by writing The Director, National Geodetic Survey, NGS Information Center C18, Rockville, Maryland 20852. Diagrams depicting such control are available by state and 1° x 2° areas. The state diagrams show only horizontal or vertical control data available from NGS, whereas the 1° x 2° diagrams (Geodetic Control Diagram) depict both the horizontal and vertical control data available from NGS, USGS, and, when available, other government organizations. Their free general information kit may be requested.

Stations have also been established by the U. S. Geological Survey. Some of their stations will not be shown among NGS data. An index map showing diagrams available can be obtained free from The Map Information Office, U. S. Geological Survey, Washington, D.C. 20242.

From the control diagrams of index maps you will be able to find what specific control data are available, how to get them and the cost.

In West Virginia the USGS in cooperation with the State, established a *Magnetic Station* in each county. Some of these are gone and some are too close to disturbing elements to obtain good declination. In the latter case, alternate points nearby in line of known true azimuth might be selected.

If government stations are damaged, or in danger of being disturbed, the establishing agency should be notified. To provide additional or updated descriptive data, it is recommended that a "Report on Condition of Survey Marker" form be submitted whenever a government marker is visited. These forms (preaddressed and requiring no postage) are available from the National Geodetic Survey.

Reading the Declination

The compass is set up over the station and the vernier set at zero. The compass is sighted on a reference mark or point for which the true azimuth is listed. Government azimuths are usually read clockwise from zero at south. The bearing of the mark is compared to the true bearing based on azimuth, and the difference is the declination. The vernier is used to read to the nearest minute. If the compass has no vernier, the declination is estimated. If the compass is sighted on true north or south, the declination is read direct.

Readings should be taken on several days at times when the needle is at its probable mean for the day (within a ½ hour of 11 AM and 6 PM EST). Use an average of these values. Do this once or twice a year for each instrument and keep a record of the observations. Show the declination of the last readings on every plat for the instrument used on that survey, giving the month, year, and specific description of where the instrument stood.

Other Ways to Determine Declination

1. Isogonic Charts of the United States are published every five years by NGS, Rockville, Maryland 20852. The latest is for 1970 and costs 60¢. A declination may be found by interpolation. The variation since 1970 is added or subtracted as shown on the chart.

2. U. S. Government quadrangles (tope sheets) show the approxi-

mate declinations for the center of each sheet for a given date; they are taken from Isogonic Charts. These must also be brought up to date by secular change data.

These two methods are approximate and should be used only as general guides. The disadvantage of using such declinations on plats to guide the retracement of lines many years hence, is that they do not represent the actual reading made on a specific instrument at a specific place that can later be found. Also they can be ½ degree or more off.

How to Calculate Variation and Declination, and Run an Omitted Call

William D. Swecker, engineer-surveyor, Elkins, W.Va., was employed to resurvey a tract with 7 calls containing about 200 acres of steep woodland. The closing call had never been run.

The first 4 calls were of a magnetic survey of 1906. These he adjusted to 1974 with a clockwise variation of 2-30 as found in the U.S. Govt. variation tables. He began at a maple (called for) and ran the first 4 calls, finding 2 more trees called for. The next 2 lines had been surveyed by the U.S. Forest Service and were described by true bearings. Swecker estimated (from a declination table) that the 1974 declination was 6 degrees west, and converted the government bearings to magnetic (by adding in the NE quadrant and subtracting in the SE). He did not rerun these two lines.

He calculated the closing call as S 65-55 W 257 poles. For convenience he ran S 66 W. At 259 poles of very rough terrain and thick vegetation with slopes up to 34 degrees, he was 8.3 feet right of the maple. Had he run S 65-55 W (with 25′ set off on his Warren-Knight Surveyor's Compass No. 130) he would have been only 2.1 feet from his objective.

Chapter 19

LOCATING TRUE NORTH

A surveyor must be able to find true north from the sun or a star. The 1947 *Manual* (p. 59) states: "Any of the various methods of Polaris observation for meridian is superior to any single observation upon the sun." *Tracy* (p. 354) says: "The simplest and most direct way of determining a true meridian is by observations on Polaris." This book describes 5 ways to find true north from Polaris.

Polaris is called the "north star" because it is the closest visible star to the true north pole. The earth rotates on an axis from west to east around the north pole. Polaris is about 52 minutes of a degree away from the pole and appears to rotate in an oval path counterclockwise around the pole.

It requires 23h. 56m. 4.09s. for the earth to make one revolution, so the star appears to do likewise. When the star is at the top of its swing, it is at upper culmination; at the bottom, lower culmination; at the west end, western elongation and at the east end, at eastern elongation. It takes about 5h. 56m. to move from UC to WE; 11 h. 58.1m. from UC to LC; 6h. 2m. from LC to EE and 5h. 56m. from EE to UC; total 23h. 56.068m. This is commonly rounded off to 23h. 56.1m.

Several companies publish an ephemeris and show methods of finding true north. The U. S. Ephemeris, a 30-page booklet, is published annually. The U. S. Ephemeris shows, among other things, the time when Polaris is at UC, WE and EE at the Greenwich meridian. These times are identical to the time when they occur on each 15° time zone meridian, provided the observer's watch is set for that time zone.

For example, suppose WE at Greenwich was at 8:40.7 PM Greenwich time. As the earth rotates about 15° of longitude each hour, WE would occur at 15° west longitude at 9:40.7 PM Greenwich time and 8:40.7 PM 15° to the west. By the same reasoning, WE occurs at 75° west longitude at 5 hours (75 divided by 15) after its WE at Greenwich, or 12:40.7 AM next day at Greenwich, but at 8:40.7 PM on 75th meridian time (eastern standard). In other words, a watch set at EST already allows for the movement of the earth for the 5 hours or 75° involved. Similarly, WE occurs at 8:40.7 PM at our Central, Mountain and Pacific standard times, *if* the watch of the observer is set at those time zones, and in which zone the observation is to be made.

Polaris at Elongation

This method may seem to be the easiest. Its main disadvantage, just as with a culmination method, is that it often occurs at bad times for observation. A form has been devised by the author for an observation

WEST VIRGINIA POLARIS OBSERVATION RECORD (Use U.S. Government Ephemeris)

Observer John M. Doe Date 3-12-71 Taken at: EE __ WE X UC __ LC __

Longitude 79°50'46"

Longitude in Decimal 79.846

Latitude 38°55'49"

Latitude in Decimal 38.93

Declination 89°08'14.62" subtracted from 90 = Polar Distance 0°51'45.38"

Polar Distance in Decimal 51.756

(Working: 60)46.00 = .77; 60)50.77 = .846; 60)49.00 = .817; 60)55.82 = .93; 60)45.38 = .756)

CALCULATION FOR TIME OF ELONGATION X OR CULMINATION ____

(Set watch by Radio - 15Mc or 7.34Mc; NBC Peacock; Local Radio Station)

Government Ephemeris time at 75th Meridian is . . . . 8:40.70 AM (PM) EST

Longitude Correction = Long (dec) 79.846 minus 75 = 4.846 x 4 = + 19.38

result is 8:60.08

Sidereal Correction = Long (dec) 79.846 x 0.011 = .878 or - .88

result is 8:59.20

For elongation, add or subtract correction for latitude (p. 22) + 0.11

Then elongation X or culmination ____ at my location is at 8:59.31 AM (PM) EST

CALCULATION FOR AZIMUTH AT ELONGATION

To interpolate cosine seconds of latitude:

lower minute cosine	.77806	lower minute =	.77806
		higher minute =	.77788
		difference =	18
subtract	15	x decimal of seconds	.82
giving cosine of latitude	.77791	to be subtracted	15

Azimuth = polar distance (decimal) / cosine of latitude = 51.756 / .77791 = 66.532 (decimal)

Azimuth in degrees, minutes & seconds 1°06'32"

(space for figuring)

.77791)51.75600 = 66.532
466746
508140
466746
413940
388955
249850
233373
164770

18 x .82 = 36, 144, 14.76

.532 x 60 = 31.920

CALCULATION OF TRUE BEARING OF LINE TO MARK (if any) to right →

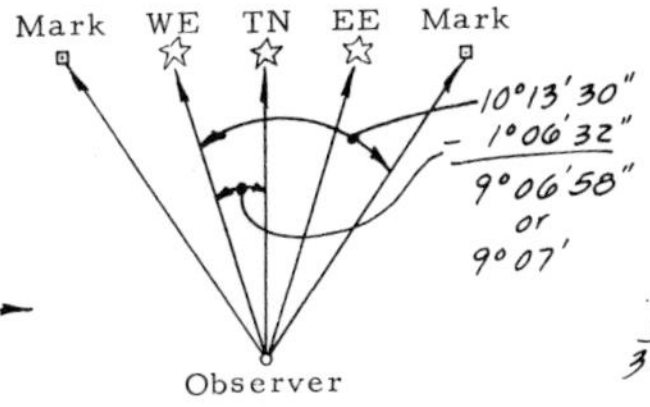

L.G. Sturgill

at elongation or culmination and is shown here filled in with real figures. It is usable in other time zones in the same manner.

Instructions for the use of this form follow:

1. Calculate your long. and lat. within a few seconds.
2. Calculate time of elongation at *your* long. thus:

A. Subtract 75 from your long. and reduce to a decimal. (Example: 79-50-46 minus 75 = 4-50-46 = 4.846.)

B. Multiply this by 4 = 19.38, and add to time of WE. (If you are *east* of 75°, subtract your long. from 75 and *subtract* the long. correction from 8:04.7.)

C. Subtract sidereal correction of 0.011 x 79.846 = 0.88. (This is the proportionate part of the 3.93m. that the earth rotates shorter than 24 clock hours.)

D. For lat. correction, interpolate from Ephemeris p. 22, column 2, and add or subtract as shown by the signs in the Ephemeris.

We now have time of WE at our location as 8:59.3 PM EST. The next job is to calculate the azimuth of Polaris at that time. The "azimuth" of Polaris is the horizontal angle between the star and true north. The form shows how as follows:

1. Insert cos. of the lower minute (lat. 38-55) in place provided.
2. Insert cos. of the next higher minute (lat. 38-56).
3. Subtract and get 18.
4. Multiply 18 by the 49s. as a decimal (0.82) and get 15.
5. Subtract 15 from the lower minute cos. and get the cos. of your lat. interpolated for the 49s.
6. Divide the polar distance as a decimal by the cos. in (5) above and get the azimuth sought = 1° 06° 32″.

The next jobs are in the field. The watch should have been set by shortwave radio (5, 7.34, 10 or 15 Mc) or by NBC-TV Peacock.

1. About 8:15 PM EST set transit over your station.
2. Clamp plates at zero and level up. Sight on an illuminated tack in hub or at some good mark visible both day and night. Clamp bottom plate and loosen top plate.
3. Turn left (in example shown) and sight on Polaris. Crosshairs must be illuminated by flashlight. At 8:59.3 PM EST star should be moving straight down and at a vertical angle equal to your lat. plus refraction (1 minute here), or 38-57.
4. Clamp the top plate and read the angle. If the mark is 10° 13′ 30″ to the right of Polaris, the true north is 1° 06′ 32″ to right of WE, then true north is 9° 06′ 50″ west or left of your mark.
5. Drop a line to the ground at 9° 06′ 50″ and mark it by hub and tack, either that night or next day.

6. Next day read some azimuths to several prominent objects for reference checks.

Note: a second observation may be taken by reversing the telescope and rotating 180° and averaging the two readings; this will increase the accuracy especially if the instrument may be out of adjustment.

Polaris at Culmination

The same form and procedure is used for sighting Polaris at culmination, except:

1. No azimuth needs to be calculated because it is always zero.
2. Insert on the form the time of *UC* from the Ephemeris. If you are observing at LC, add 11h. and 58.1m. to the UC time.
3. Omit the latitude correction.
4. Polaris is found at a vertical angle equal to your lat. plus the polar distance for UC, or minus polar distance for LC; plus refraction.
5. Watch time must be carefully set or closely known because the star is moving rapidly at right angle to your line of sight. Within a few seconds clamp the plates (which were set at zero) and drop a line to the tack and hub on the ground. Or loosen the top plate and read to several prominent objects as before.

Polaris at Any Time (Any-Hour Angle)

It is often inconvenient or weather is bad at time of elongation or culmination. Therefore, various systems have been devised to observe Polaris at any convenient moment, and to calculate the angle between the star at that time and true north. The method here presented is just as easy or easier than at elongation or culmination. The surveyor therefore should use this method almost entirely.

The first job is to set up over the station; clamp the plates at zero and level the transit. Point on Polaris and clamp the bottom plate. Record the correct time of clamping within a few seconds. Drop a line to a tack and hub on the ground, or loosen top plate and read to some prominent objects, and record the azimuths. Remove the transit and proceed to the office.

The author has devised a form and instructions for an Any-Hour Angle observation. An actual example is shown here. The azimuth is *always* calculated from *UC*. Remember that UC at Greenwich time is the same as UC at your watch time meridian.

Here are instructions for use of this form:

1. Insert date of observation, longitude and latitude.
2. Convert longitude and latitude to decimals.
3. Insert declination of star from Ephemeris.
4. Determine if star was east or west of north meridian when ob-

WEST VIRGINIA POLARIS OBSERVATION RECORD at ANY-HOUR ANGLE (US Ephemeris)

Date of observation 5-16-73 Longitude 79-50-46 In decimal 79.846

Latitude 38-55-49 In decimal 38.93 Declination 89-08-29.28

IF STAR IS WEST OF MERIDIAN & obsn. is in PM X AM___ & UC in PM___ AM X
(for AM obsn. & PM UC, use UC for preceding day)

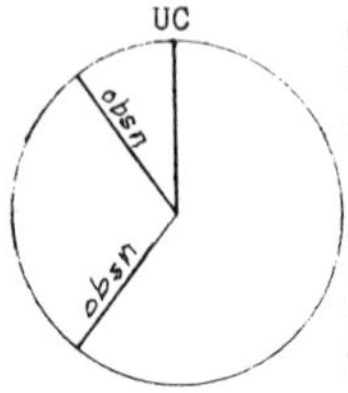

Time of observation (corrected for watch error) 8:26.00 EST
Minus long. correction (long - 75 x 4) of - 19.38
Gives Local Mean Time of obsn on 75th meridian 8:06.62
(add 12 if this is less than UC below) 20:06.62

To calculate UC at 75th meridian:
From UC at Greenwich, which is 10:28.90
Subtract sidereal (long x 0.011) - 0.88
Then from LMT above, subtract 10:28.02 = 10:28.02
Giving a west HOUR ANGLE of 9:38.60

IF STAR IS EAST OF MERIDIAN & obsn. is in PM___ AM___ & UC in PM___ AM___
(for PM obsn. & AM UC, use UC for following day)

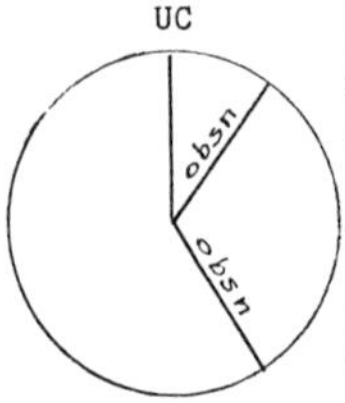

To calculate UC on 75th meridian:
From UC at Greenwich, which is ________
Subtract sidereal (long x 0.011) - ______
Giving Local Mean Time of UC of ________ = ________
(add 12 if this is less than LMT below) ________

Time of observation (corrected for watch error) ________
Minus long. correction (long-75x4) - ______
Gives Local Mean Time of obsn on 75th ________ = ________
Which subtracted from LMT of UC above
Gives an east HOUR ANGLE of ________

HOUR ANGLE CALCULATION FOR AZIMUTH (interpolate for all missing figures)

		lower 2° lat.	Observer lat. in decimal	higher 2° lat
		38-00	38.93	40-00
			+% 46.5	
High hour angle*	9:38.40	37.00		38.00
Observed hourangle	9:38.60	36.95	37.42	37.95
Low hour angle*	9:48.40	34.70		35.60
(subtract low from high)		2.30		2.40

Add or subtract correction for declination 0.11
Then final AZIMUTH (in decimal) is: 37.53'
which converted to minutes and seconds is: 0°37'32"

*see Ephemeris page 18 or 19

48.40 .98 .98 37.95
38.60 23 2.4 - 36.95
980 294 392 1.00
196 196 x .465
2.254 2.352 .465
+ 34.70 + 35.60 + 36.95
36.95 37.95 37.415

Observer Sipe

Form by F.Henry Sipe LLS
Elkins, W.Va. 8/3/71

served. Do this by comparing time of observation with time of culmination. If star was west of north, use first set of data; if east, use second set.

5. Insert actual time of observation, corrected for watch error.

6. Subtract correction for longitude, (longitude minus 75) times 4. (Add if you are east of your time zone meridian.)

7. Insert time of *upper* culmination from Ephemeris, without any adjustment.

8. Subtract sidereal. This allows for prorated part of the 3.93 min. shortage in the so-called 24-hour rotation of the earth.

9. Add 12 if needed to allow mathematical subtraction.

10. Make final subtraction and get the *Hour-angle.*

11. Insert *your* latitude, and figure what % of the way it is from the lower latitude toward the higher latitude (by interpolation).

12. Insert next lower and higher even-numbered latitude that encompasses *your* latitude.

13. Insert the hour-angles and latitudes, from the Ephemeris.

14. Insert azimuths for those hour-angles and latitudes, from Ephemeris.

15. Subtract low from high azimuth as an interpolation base.

16. Interpolate to obtain azimuth at *your* hour-angle and latitude.

17. Insert interpolated correction for star's declination, and add or subtract.

18. Result is the azimuth at time of *your* observation.

The next day set up over your station, clamp plates at zero and level. If a tack and hub were set directly under Polaris, sight on it, clamp bottom plate, loosen top plate, and turn azimuth angle (0-38 to the right) and you are sighting on true north. If you sighted on another mark the previous night, turn off the proper angle as shown in the elongation example. For example, if your night mark was 45-28 to right of the star, true north is 45-28 minus 0-38 or 44-50 to left of the mark.

Polaris Observation in Daylight

The north star can be seen nearly an hour before official sunset. This greatly enlarges the available hours for convenient observation and eliminates the need for illuminating the crosshairs. The surveyor should practice focusing on stars and mark the focusing position on the scope so it can be turned to without looking at a star.

To find Polaris, level the transit with plates at zero. Set the vertical angle at the latitude of the place of observation plus refraction. If the lat. is not known, it may be found by locating your position on a tope sheet. Use a ruler of appropriate scale for the sheet involved. By proportionate measurement or interpolation find the lat. to the nearest several seconds. The lat. will approximate the altitude or vertical angle of the star at elongation.

The star however, will not likely be at elongation. Its distance above or below the pole will be about as shown in the following table made at Elkins at lat. 38-55.8:

Minutes of time before elongation	*Approximate hor. minutes of degree from elongation*	*Altitude before elongation**		
		West	*East*	
180	22.0	39-35	38-17	*To these figures, add 1′ to cover refraction
150	14.0	39-30	38-22	
120	9.6	39-23	38-29	
90	5.6	39-17	38-35	
60	2.4	39-10	38-42	
50	1.7	39-08	38-44	
40	1.2	39-05	38-47	
30	0.8	39-03	38-49	
20	0.5	39-00½	38-51½	
10	0.3	38-58½	38-53½	
5	0.1	38-57	38-55	
0	0.0	38-56	38-56	

If the star is outside the limits of this table, the observer will need to increase the range of search. Obviously the star must be less than the polar distance away from the elongation altitude. (Polar distance is 90° minus the star's declination). Patience is required for the search. Search from side to side changing altitude as needed.

Finding True North Without Compass or Transit

Many textbooks state that a true meridian can be found by lining up two plumb lines with Polaris at elongation, and measuring the azimuth offset. Or by substituting some other south sight that can be moved laterally. The details of this method and the accuracy that can be obtained have not been well-described. The writer therefore conducted experiments to test methods and learn what accuracy was obtainable.

Two 2x2 hubs with tacks were set 50 feet apart in a true north line established with a transit by a conventional method. A 20-foot tripod was set over the north tack, and a plumb line suspended from it. The tripod was made with three 1½″ aluminum tubes at the bottom, then three round wood rods hinged at the top; on one wood rod a 12-foot length of smaller aluminum tubing was bolted. A 6-foot tripod was set up 20 feet southerly. A calm night is necessary for best results.

After experimenting with various kinds of lines and devices, it was found that a 75-lb. test white nylon line (about 0.05″ diameter) worked best for the high tripod, and a size D yellow nylon thread for the low tripod. Both lines were illuminated by electric light about 15 feet southwest of the south tripod. A piece of wood 1x2x10″ was nailed to the ground at the south 20-foot point in an east-west direction, and true south marked on it. A white strip of cardboard was then thumbtacked to the wood strip; the low tripod and plumb line was set over it.

Time of elongation was then calculated. About ½ hour before elongation the south line was aligned with the north plumb line and Polaris. At 5-minute intervals, a mark was made on the white card under the south plumb bob. The azimuth of each pencil mark from true south was calculated and compared with the azimuth at each corresponding time of observation. Calculations can be shortened by previously marking 1-minute azimuth lines on the white card.

The first observations were 3 to 4 minutes in error. Practice and improved techniques soon reduced the error to 1 minute, and then closer than ½ minute, with an occasional hit "on the nose". The average of several observations, whether taken at elongation, culmination or any hour-angle, can keep the accuracy to within one minute.

A good technique is to sit on a box or stool just south of the south plumb line, close one eye and place the other within an inch of the yellow line. This closeness allows the north plumb line and star to be seen *through* the yellow line. One hand on a tripod leg can make fine adjustments of the south plumb line.

Because an error of 0.006 foot will make a difference of one minute in azimuth in 20 feet, all observations should be made with a precision that will keep total error to less than that amount. The position of the north plumb line over the tack, and the time of observation if at other than elongation are especially critical.

The technique above described is valuable to any surveyor who surveys only with an open sight compass. The accuracy obtainable is superior to the average solar observation with a transit.

Equal Altitudes of Selected Stars

A method that requires the observer to arise about 4 AM, is called Equal Altitudes. If the surveyor fixes the direction of a line to a certain star about 8 PM, at a certain height above the horizontal, when it is west of N and is descending, then finds the direction of the same star about 4 AM, after it has sunk below the horizon and risen again on the east side of N to the same altitude (height), and divides the angle between the 2 lines, the middle will be the true N.

In daylight establish a stake and tack station (A), away from lights and local attraction. Set another stake (B) and tack (B) in a northerly direction several hundred feet away, visible from the first station. Or for the second station, select a prominent point that can be seen both day and night. After dark set up the transit over station (A), clamp the plates together at zero, carefully level the telescope, and with a light on (B) and on the crosshairs, point to (B). Select a prominent star which is descending SW of N (make a diagram in the notes to show which star). Loosen the top plate and point the telescope toward this star. Clamp the vertical arc at the nearest even degree below the star. Clamp the top plate and follow the star with the slow motion tangent-screw and vertical hair as the star moves SE. When the star is at the crosshairs, record the time, vertical angle, and horizontal angle. Loosen the vertical arc, set it at the next lower degree or half degree, and follow the star as before till it hits the crosshairs; record the time, vertical angle and horizontal angle. Repeat the process for the next degree or half degree down.

Bring the transit inside. About ½ hour before the AM observation, set up the transit over (A), level, and clamp plates at zero. Point at (B), release the top plate and set the vertical arc on the last vertical angle of the night before. Pick out the right star and follow it with the vertical hair as before until the star is at the crosshairs. Record the time, vertical angle, and horizontal angle. Now set the vertical arc at the second altitude of the night before and follow the star as before. When it reaches the crosshairs, record the time, vertical angle, and horizontal angle. Repeat for the third reading. For three such readings, the notes could appear thus:

Sept. 15, 1964, readings on end star of Big Dipper handle (station (B) is airplane beacon).

Time	*Vert. Angle*	*Hor. Angle*	*Time*	*Vert. Angle*	*Hor. Angle*
8:32 PM	28-00	329-14	4:27 AM	28-00	57-39½
8:37½ PM	27-30	329-30½	4:21½ AM	27-30	57-23
8:42 PM	27-00	329-49½	4:17 AM	27-00	57-00

The midpoint of each set of readings is calculated thus: 360-00 minus 329-14 is 30-46, + 57-39½ is 88-25½, divided by 2 = 44-12.75, which subtracted from 57-39.5 is 13-26.75. The same calculation is made for the other 2 sets of readings; add the results together and divide by 3.

Next day set the transit over (A), level, clamp plates at zero and point on (B). Loosen the top plate, turn to 13-27 (or 13-26.75) and clamp the plate. Lower the telescope to the ground, set a stake and tack in line; *this is the true north.*

The time to make both observations can be found on a previous night by recording with clinometer the vertical angle and time of sightings on several stars at night and in the early morning. The method can be varied by experience.

This method is not recommended in cloud country because the star selected in the evening may be clouded over in the morning. Also temperature is often unfavorable in early morning hours.

Converting From True to Magnetic and Vice Versa

Surveyors occasionally must convert to bearing of a given line on the ground, or a random line to be run, from true to magnetic or vice versa. This calculation is often misunderstood. The rule is:

When Declination is West

From TRUE to MAGNETIC: in NE and SW quadrants, add declination to the true bearing. For example: with a 6 degree west declination (as in West Virginia), N 60 E true becomes N 66 E magnetic. In the NW and SE quadrants, subtract declination from the true bearing. For example: S 60 E true becomes S 54 E magnetic.

From MAGNETIC to TRUE, the above signs are just the opposite.

When Declination is East

The calculations and signs are just the opposite from the above. For example: with a 17 degree east declination (as at San Francisco), a true bearing of N 60 E becomes N 43 E magnetic, and a S 60 E true bearing becomes S 77 E magnetic.

A paper diagram showing both arrows and the line direction will confirm the calculation. But remember, the line ON THE GROUND does not change its location. And do not confuse these declination changes with changes in VARIATION.

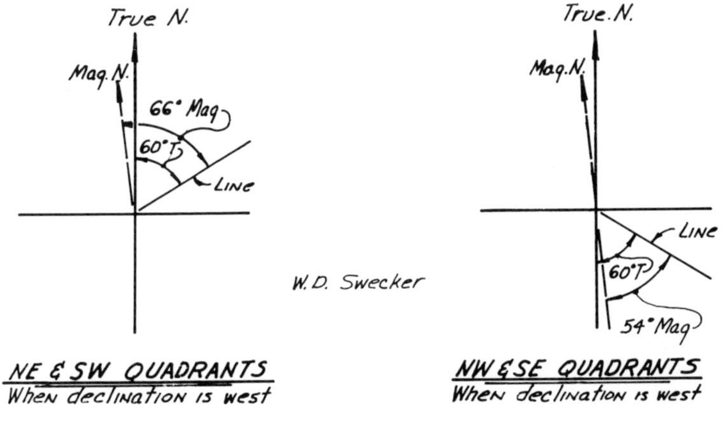

Chapter 20

HOW TO SURVEY A RURAL PROPERTY

1. Discuss with client: deeds, ground evidence, adjoiners, contract.
2. Visit Assessor's office: Obtain tax map and card record data.
3. Visit County Clerk's office: abstract deeds of clients and adjoiners. Find calls and go back to original or to common grantor.
4. Plot calls: check for errors; match adjoiners; do not run Lats. and Deps.; list problems.
5. Calculate variations since origin of calls.
6. Visit ground with client and adjoiners; look for and flag markers.
7. Talk to adjoiners; inform them of plans; obtain permission to go on their property; try to get statement on corner locations or that they do not know.
8. Decide on methods and equipment best adapted for the job.
9. Run trial lines.
10. Office work: reduce slope distances to horizontals; calculate bearings and distances between markers; plot random survey; calculate error of closure.
11. Decide on corner locations; make preliminary conclusions.
12. Send report and plat to client and adjoiners for comment.
13. Evaluate replies; make such adjustments as are needed.
14. Make final decisions as to corner locations, etc.
15. Final calculations: bearings and distances; balance Lats. and Deps.; calculate area; seldom balance bearings and distances.
16. Final plat: make work plat first, then copy it for the original.
17. Write description and report.
18. Offset, mark and paint lines; establish monuments.
19. Furnish report, description and plat to client; keep original plat. Copy of plat to adjoiners? If payment is questionable, consider withholding data until payment is made.
20. Disputed lines: see next chapter.

Note

One of the most serious mistakes made by land surveyors is not tieing to the nearest corners of the parent tract. For example, a parcel is to be sold from the edge of a larger tract, three lines being interior and new; the fourth line adjoins another landowner. The "outside" line must usually be established by tieing to the nearest corner in each direction. If a fence is along this line, it cannot be assumed to be at the correct line location, unless a line agreement of some kind is in effect.

Chapter 21

PRELIMINARY SURVEY PROCEDURES

The previous chapter outlined the general steps in making a rural land survey. This chapter begins the details of the work. There is much to be done before starting the actual field survey. Sometimes a third of the entire job must be devoted to checking records, plotting, etc. This and the following chapters give in chronological sequence, how a survey is made.

Discussion with Client

Many landowners know little about what is required to make a good survey of their property. They must trust the surveyor to do what is proper. He must diagnose the troubles and explain to the landowner what needs to be done—what constitutes a good survey. Adjoiners deeds must be checked, good markers set, lines marked, and so on. These explanations can begin at the first interview and continue throughout the job. The following items should be used as a check list of data that may need to be known about the job.

The Client

1. Name, including spouse, as they appear in deeds.
2. Mail address and phone number and how to reach on short notice.
3. Occupation or place of work and phone.

Purpose or Reason for the Survey

Discuss why the client wants a survey. This should help determine *if* there is a need for the survey or if the problem can be solved in some other way. The surveyor can also know if he is qualified to do the job, has time to do it, and has proper tools for the work. If the boundaries are reasonably well defined by roads, streams, fences, etc. a resurvey may not be needed. But if timber is to be cut on either side of a vague line, if a dispute has developed, or if both adjoiners wish to redefine the location, a resurvey of all or a part of the property may be required.

The Source of Title

1. Deed—original or copy, or a reference to the recorded deed book and page number, county and district; name of grantors and grantees; year of conveyance and acreage.
2. Other data—will, mortgage, deed of trust, contract, court decree, boundary agreements, previous surveys, plats, leases, former owners, abstracts of title.

3. If the land was inherited without a will (intestate), obtain the title reference of the prior owner.

Adjoiners

Obtain the names and addresses of adjoiners, including spouses. Note these on a rough sketch of the area, which the surveyor may make during the first visit with the client. The client must understand that the rights and calls of the *adjoiners must be given fair consideration* in the survey. If laps are found, they must be surveyed, mapped, and perhaps referred to an attorney for an opinion on who has the better title.

Policy in Disputes with Adjoiners.

1. Obtain copy of adjoiner's title instrument; go over it with him; compare with recorded instrument and evidence on the ground; *go back to common grantor.* Be friendly
2. Obtain permission to use his land for trial surveys
3. Give due weight to his title evidence
4. Keep him informed of progress, but use discretion
5. Inform him of your tentative conclusions
6. If he disagrees, study his additional evidence, if any
7. Furnish adjoiner with plat and statement of reasons for your position
8. Adjoiner should be encouraged to clearly accept or rebut your evidence, rather than leaving a vague atmosphere
9. Prepare boundary line agreement, if desirable, or state in description of survey "thence with an agreed line xxx". Acquiescing parties could sign the plat

Agreement with Client

1. What is to be done by surveyor
   a. Abstracting and plotting
   b. Running lines—random and true
   c. Marking lines
   d. Painting lines
   e. Monumenting corners
   f. The plat—number of copies and disposition
   g. Description of survey—number of copies and distribution
   h. Certificate of survey—number of copies and distribution
   i. Recording the plat and description. Insist that the plat be recorded, either as part of a deed or as a separate instrument. A recorded plat (a) strengthens the owner's title, (b) gives notice to adjoiners, (c) provides data for other surveyors to build on, (d) provides data for public use, and (e) reduces future survey costs

j. Information to be given to, or discussed with, adjoiners
k. Discuss boundary disputes
l. Locate easements?

2. Helpers—who is to furnish and pay
3. Charges for the Work
    a. Rate per hour (job, travel, helpers)
    b. Mileage rate—one or both ways
    c. Markers—type, who to furnish
    d. Plats
    e. Recording fees
    f. Miscellaneous (aerial photos, paint, etc.)
    g. Who will pay the surveyor and when
4. *Complete Date*

5. *A Written Agreement.* Some states require a written agreement between the surveyor and client. If the matters in this chapter are fully understood by the client, no written agreement appears necessary. However, if an agreement is desired, the following sample is offered:

> I, ________ (client) of ________ (address) do authorize ________ (surveyor) to ________ (describe the job) for the following consideration at the rate of ____ per hour (or for a lump sum) for a ____ man crew. The client agrees to furnish ____ helpers. Time includes travel between ________ and the job. Mileage will be charged at the rate of ____ cents per mile between ________ and the job. Other items are to be paid by the client at cost, such as corner markers and paint. Office work shall be paid at the rate of ____ per hour. Minimum fee, if any, is ____. The surveyor hereby agrees to make the survey in accord with the applicable laws and regulations and by proper methods suitable for the job.

The agreement should be dated, signed by the client, surveyor, and two witnesses, and a copy made for each party.

Ground Conditions

1. Markers present, type, location
2. Marked lines, fences, and their history
3. Adjoiners
    a. Names, including spouses
    b. Addresses or places of residence
    c. Approximate acres, year acquired and how
    d. Acquired from whom
    e. Any line disputes
    f. Adjoins which lines of client (make sketch)
4. Easements—electric, telephone, pipe lines, roads
5. Timber cutting, when and by whom
6. Other possession evidence

Review of the Records

A study of the records is next. In most instances, deed calls form the basis for the survey. Deeds are recorded in the office of a County Clerk or Recorder. That office maintains an alphabetical index of grantors and grantees of deeds, also indexes for wills, deeds of trust, and other instruments. Indexing systems vary between counties; a clerk will gladly explain the system. Assessors' maps and aerial photos should be checked.

Abstracting the Deed. An original deed should be compared with the recorded deed. In former years errors occurred in copying; in recent years photo copying eliminates such errors. If there is no original, abstract the recorded deed. If it does not contain calls (bearings and distances), follow the chain of deeds back (or forward) to a deed that does. If no deed with calls is found, the land will likely be described as part of another tract. The surveyor should seek calls for the larger tract and calls for outsales. If he is lucky to find both, the problem may be solved. If no calls can be found, corners may have to be agreed upon with the adjoiners. The following form of abstract is satisfactory (note abbreviations):

John R. Jones & Hazel, wife	Deed dated 3/23/1897
to	D. B. 54, p. 581
George S. Roberts	Randolph Co. W. Va.

Conveys: 43 acres on L. Fk. of Buckhannon River in Middle Fork District in said county.

Description: Beg. at a stake (st.) and pointers (ptrs), corner (cor) to Fred Balli, and with said (sd) Balli

N 69 E 178 poles (p) to st. & ptrs, cor. to Fred Balli, and with (w/) sd Balli

(etc.)

S 10½ E 35 to beg. cont. 43 acres, m or l.

Subject to easement for a RR right of way (RR r/w);
Reserving all coal, oil, gas, and other minerals (COG et al.);
Excepting 12 acres conveyed (conv) by —— to —— by deed dated ——, DB —— p. ——.
Being part of (BPO) tract of 164 acres (ac) conv. to grantor by Harry Jones, by deed of 1/16/1872, DB ——, p. —— in sd. Co.

The surveyor can use other abbreviations of his own. Deed references are often written 54/581 or 54-581. Calls should always be written *in tabular form,* starting a new line with each new call; this makes it easier to locate, plot, and compare calls. Some items in the above example may be deleted, such as "dated", DB, p, W. Va. The term "more or less" can be omitted.

The deeds of adjoiners are also abstracted. If deeds are found that seem to match the client's deed lines, the search may stop there temporarily. If the adjoiner is a public road, railroad or large stream, it may be necessary to look for and copy plats, photos, or easements. A description of corners and lines of National Forest land is found in USFS offices, in addition to County records. Boundaries of modern public highways are usually found in plats and deeds in the Clerk's office and in the office of the State District Engineer. Some secondary roads have no instruments of record, but there may be a record in the County Court minutes. Most railroads were built many years ago and records may be hard to find; railroad offices may contain helpful data.

Other Instruments. Most of the comments under deeds apply to the search for other instruments. Record of lawsuits over title or location will be indexed and found in Clerk's offices. If a deed was made as a result of a Circuit or higher court decree, it will usually be found in the County Clerk's office. Maps are often found in the higher Clerk's files for a specific case. The County Clerk's office may also contain special records and indexes of maps and surveys.

Plotting Calls

Make a rough sketch of the calls after the abstract of a deed has been made. This gives a bird's-eye view of the situation and helps match adjoiners' deeds. However, it is on the drafting table that the main plotting begins. The scale of the plat is important; it will be affected by one or more of these factors:

1. Space for the amount of detail desired
2. Size to compare with other maps or aerial photos
3. Adequate space on the drafting table
4. Size of paper available
5. Size of printing machine used to record or make copies

Use 8½″ x 11″ paper, if practicable, so as to fit normal correspondence files. However, paper 30″-36″ wide is sometimes necessary for preliminary plats. Paper will usually be oriented with the four edges in the four cardinal directions (N, E, S, W), with N at the top. If some other position is necessary, a N line (or E-W) should be shown on the paper. Select the beginning point so as to get all the tract on the sheet. A hard pencil (6H to 8H) will help accuracy.

Finding Errors. Many problems arise in plotting calls. Most old deed calls fail to close. Large errors of closure are often caused by omitting calls, by using SE instead of SW, or using a reversed call—SE instead of NW. Errors may be caused in copying, but more often because writers of deeds have combined calls in deeds or surveys made at different times. Copying errors can be corrected by comparing the same call in several deeds.

Aerial Photos. A good way to find large errors—and sometimes smaller ones—is to compare the plat with an aerial photo. Photos may be inspected in most offices of the Soil Conservation Service (SCS) and the U. S. Forest Service. Sometimes the Agricultural Stabilization and Conservation Service (ASCS or ACP) has later flights. The scale of all photos, particularly enlargements, must be checked before using them. Photos are made from airplanes several miles in the air, flying in a N-S or E-W direction. The photos overlap, so that with a stereoscope the third dimension (height or depth) is visible.

Thus a deed calling for the meanders of a ridge, stream, or road, or anything identifiable on the photo, can be easily located, and errors allowed for. Photos taken when leaves are off are most helpful. Fence rows, edges of woods, and edges of timber of differing ages can often be seen. The type of terrain, difficulties of field work, and travel routes can be noted.

"Contact" photos are about 3″ per mile—roughly 100 poles per inch. However, the scale varies between photos and on different parts of the same photo, so it is best to plot the calls on the scale closest to the part of the photo where the tract is located. Select a photo that has the tract near its center because the outer limits of a photo contain distortion. A table has been prepared for converting poles to inches for various photo scales. Use the table scale (A to F) closest to the photo scale.

TABLE FOR PLOTTING POLES IN INCHES FOR SIX AERIAL PHOTO SCALES

Move decimals and add as needed. For example, 79 poles in column A = 69.3 + 8.9 = 78.2 or 0.78 inch.

	A	*B*	*C*	*D*	*E*	*F*
R/F	20000	20500	21000	21500	22000	22500
ch/inch	25.25	25.88	26.51	27.14	27.77	28.40
poles/in.	101.00	103.52	106.04	108.56	111.08	113.60
poles						
1	.99	.97	.94	.92	.90	.88
2	1.98	1.93	1.89	1.84	1.80	1.76
3	2.97	2.90	2.83	2.76	2.70	2.64
4	3.96	3.86	3.77	3.68	3.60	3.52
5	4.95	4.83	4.71	4.60	4.50	4.40
6	5.94	5.79	5.66	5.53	5.40	5.28
7	6.93	6.76	6.60	6.45	6.30	6.16
8	7.92	7.73	7.54	7.37	7.20	7.04
9	8.91	8.69	8.49	8.29	8.10	7.92
10	9.90	9.66	9.43	9.21	9.00	8.80

The use of calculators has made such tables less used. For example, to convert poles to inches at a scale of 113 poles per inch, insert 0.00885 in the calculator and multiply by the poles in each call.

The scale of a photo is found in this way: on a U. S. Geological Survey quadrangle map find two points a mile or so apart that can be found on the photo. Multiply the map inches by 315.7 (if its scale is 1:62500); divide the product by the photo inches between the same points. (If the map scale is 1:24,000, multiply by 121.21.) The result will be the scale of that part of the photo in poles per inch. It is good to average several lines at right angles to each other. All parcels can be plotted on photo-scale, shifting to a standard scale later.

Changing Map Scales. The surveyor must sometimes reduce a large map to a smaller size. This can be done with a pantograph or by replotting calls. Another way is to plat proportionate distances. In this method, an 8½ x 11 sheet of tracing paper is taped over the large map so that measurements can radiate out from a certain point on each map. The reduction proportion should be worked out in advance. If it is a simple 4 to 1 reduction, use a 10-scale to measure from the point to each corner on the large map; then measure off the same distance on the small map with a 40-scale. Connect the points so located on the small map just as they are on the large one. If the reduction is not even, the factor can be set up on a calculator, and each reduction quickly figured.

A Visit to the Area

Before surveying large tracts, a trip to the land, preferably with the client, is usually desirable. This is to:

1. Locate one or more authentic corners or other evidence
2. See how routes of travel, rivers, cliffs, fences or other obstacles will affect the work
3. Discuss with (write to) adjoiners these items:
    a. The reason for the resurvey
    b. Corner evidence
    c. The meaning of random lines
    d. Reports of progress or final results
    e. The use of their land for access, etc.
4. Look for signs of possession

Weighing the Evidence

The surveyor analyzes the evidence, and decides, on the basis of laws of surveys and court decisions, what weight is to be given to each fact, and how this will guide the resurvey. These decisions are subject to revision as new evidence is discovered. Most disagreements between surveyors arise from failure to find all available evidence.

Chapter 22

THE RANDOM (TRIAL) SURVEY

Orientation of the Survey

Before field work begins, the surveyor should know on what meridian to orient his survey. If it is to be made from the true meridian, the answer is clear: a celestial observation must be made or the meridian must be brought in from a survey or lines of known true bearing. One of the most important steps in upgrading land surveys would be to require all surveys to be made from the true meridian.

However, most rural surveys in West Virginia and likely in other eastern states, are still made by the magnetic meridian. In fact, a West Virginia law originating in Virginia in 1849 still provides that surveys be made by the magnetic meridian. (See Chapter 2.) The law does provide that the declination be shown too, but this means only an approximate tie to the true meridian. With knowledge and care it is possible to use the magnetic meridian effectively.

Many surveyors believe they must use the magnetic bearing of some former call of the tract being surveyed, or the bearing of an adjoiner's line, as their orientation base. If that practice was carried out, orientation would soon be hopelessly entangled with meaningless meridians impossible to understand. Index errors and carelessness of surveyors, landowners and attorneys would result in great confusion.

The proper way to orient a transit magnetic survey is to set the instrument on a station of the survey, clamp the plates or vernier at zero with no declination set off. Lower the needle and sight on magnetic north. Clamp the bottom plate and loosen the top plate and then turn to the line or bearing being run; use this bearing as the orientation base. The plates may also be read.

The procedure with a compass is about the same: clamp plates and vernier if any, at zero and sight on the bearing or line to be run. The tie to true north has been described in the chapter on declination. The amount, date and place the declination was read must be shown on the plat. This provides the link for the surveyor of the future to get a start on retracing your lines.

Preparation

Field work is started after obtaining all or most of the data needed. List on a notebook sheet, in numerical order, each job to be done, in this manner:

1. Beg. at stone pile at NW cor. 146/48 deed.
2. Run N 7 E, at 3 ch. look for beech cor. (Jones 62/73)

3. Tie to fence cor. at field at about 26 ch.
4. Look for sugar cor. at 152 p. (38.0 ch.)

The bearing of random lines is adjusted for variation. Most older deeds in West Virginia have the distances described in poles; convert these to chains for measuring with a 2-ch. tape. If there is much converting of feet, tables can be made. Example: calculate the pole value of 1 to 10 feet; the poles for any footage can be added by setting over decimal points thus: 1 ft. = .0606 p. and 100 ft. = 6.06 p. Or set 0.0606 in a calculator and multiply each call by it.

Determine the size of crew needed. For some compass surveys, the crew can be reduced to 2 men—a compassman and a chainman. If there is brush to be cut, consider a third or fourth man. Select lines for running, if feasible, so that return foot travel to the car is as short as possible. Use the check list to see that all items are packed in the vehicle. The author's check list is shown.

Survey Equipment Check List

Transit/Compass
Tripods
Tape/Stadia
Clinometer
Notebook
Flagging
Pencil/string
Folder, maps, deeds
Photo & carrier
Rods/level
Nails, caps, tacks
Keel/stakes
Traverse tables
Slope reductions
Variation curves
Ruler
Magnifier
Axe, hand
Chisel
Trowel
Canteen
First aid kits
Insect repellant
Watch & knife
Bandana
Toilet paper
Paint/brush
Scraper
Protractor
Vest
Non-mag. Glasses
Clippers
Hard hat
Lunch
Knapsack

Field Notes

Notes must be complete, clear, neat, and honest. A looseleaf book eliminates the need for indexing and allows notes to be filed in the case folder but watch for magnetic disturbance. Number each page in the upper outside corner. Start each day with a new page. Cross reference as needed. On large surveys allow several pages for indexing pages to corners or use a key sketch. Do not erase; draw a line through the error and write the correct data above. When later notes are added to a previous page, be sure to date the addition. Use a zero in front of decimals for numbers less than one; for example 0.63. Record name, address, Social Security number, job assignment, weather, and time in the upper right page.

Surveying With 2-man Crew

The beginning should be at the most authentic identified corner of the tract. It need not be the beginning corner of the deed. The line should be run in the same direction as the deed calls, all else being equal. If there is a choice, run a line away from the sun, to improve the compassman's visibility. On a N-S line this eliminates the need for the surveyor to move halfway around the compass to check the N end of the needle.

A good description of the starting point should be recorded at the top of the first left hand page of the notes. This point may be referenced now if desired. The compass is set over the corner, levelled, and the random bearing pointed at and recorded. If the needle is not shimmying normally, the tripod should be tapped lightly to free the needle.

Offsets. If the corner is a tree or an object that cannot be set on, the compass is placed about two feet away, in the random line of sight; this usually involves shifting the compass and tripod until a backsight strikes the middle of the tree or object. If the beginning is at a wire fence, or if the line hits a fence or other obstruction, the random line will have to be offset to avoid wire fence attraction, or get away from other obstacles. To make such an offset, measure horizontally at right angle to the random line bearing and set a small stake. The correct angle can be obtained by trial and error with the compass. The offset is recorded as a surveyed line. For offset distances, see chapter on magnetic variation.

Running the Line. From the new starting point, the compass is turned to the random bearing. The chainman should pass the zero end of the tape around some tree or object so it cannot rub on the tripod leg as it is being pulled forward. The compassman points out the line of sight to the chainman, who should stand close to the sight line and get the same "picture" as the compassman. He points out some tree or object in or near the line about 2 ch. away. The chainman, with tape, rod, and axe, proceeds along the line. The tape is soon retrieved by the compassman, who allows it to slip through his hand. If the chainman gets too far from the line, the chain is tightened to stop the chainman until he gets back on line. When the 2-ch. point nears the compass, the compassman tightens his grip on the tape and calls "chain". This is the signal for the chainman to stop, turn, face the compass, lay down the axe and raise the rod. The chain is held in one hand down at arm's length (about the knee) with the zero point at the middle of the rod which is held vertically in front of the center of the chainman. The compassman holds the 2-ch. mark over the station at about the same height. With a 15 lb. pull on the chain, the rod and chain are moved to the right or left as directed by the compassman. The procedure is one of measuring distance and direction alternately, then together. A check may be made with both ends of tape on the ground. Make final check with both men standing "sideways" to the line and facing the tape to better check the tape ends being vertically over the station.

When the surveyor is satisfied with the front rod's location, he lets the tape fall; this is a signal for the rodman to stand erect, close to and behind the rod, while the compassman *carefully* reads the vertical angle with a clinometer or Abney level to the nearest ½° (to ¼° on steep slopes). The clinometer is sighted on a point on the rod or chainman previously determined to be about eye level of the compassman. Record the reading along with bearing and distance; the surveyor then signals OK by voice or arms to the chainman. The front rodman may also have a clinometer to read on the compassman and call his figure to the compassman for averaging the two readings.

The chainman *marks the spot* on the ground, lays the rod aside, and cuts a stake. The stake should be about 1½″ in diameter and 3 ft. long, sharpened at one end and shaved flat at the other. Drive the stake securely; if it hits a rock the chainman can pile stone around it or offset it *in the line of sight* one or two links either way (advise the compassman of such offsets). A nail and bottle cap can be used if stakes are in the way or likely to get pulled up. If anything other than a stake is used for a station, it should be described in the notes. Presawed stakes or lath may be used.

Meanwhile the compassman raises the needle off the pivot, folds the sight vanes and brings the equipment forward to the new station. En route he watches for old marks of a prior survey and for signs of possession. If something of interest is found, it is measured to and described in the notes on the right hand page opposite the proper distance of the left hand page. Measurement is usually made at a right angle to the line of sight. When running along a fence or stream, the right angle distance thereto from each station should be recorded.

For Stations 2 Chains Horizontal. The previous system will provide stations every 2 ch. *slope* distance; if stations are to be every 2 ch. *horizontal,* a different method is used. The 2-ch. slope distance must be extended a specific number of links according to the steepness of the slope. The extra links for various slopes to give 2 ch. Hor. Dist. are shown in the following table:

%	*Degree*	*Links*	%	*Degree*	*Links*	%	*Degree*	*Links*
5	3	0.2		12	4.5	40	21.75	15.4
	4	0.5		13	5.2		22	15.7
	5	0.8	25	14	6.1		23	17.3
10	5½	1.0		15	7.0		24	18.9
	6	1.1		16	8.0	45	24¼	19.3
	7	1.5	30	16.75	8.8		25	20.6
	8	1.9		17	9.1		26	22.5
15	8½	2.2		18	10.3	50	26½	23.6
	9	2.5		19	11.5		27	24.5
	10	3.0	35	19¼	12.0		28	26.5
	11	3.8		20	12.9	55	28.75	28.3
20	11¼	4.0		21	14.2		29	28.7

For a slope not in the table, multiply the exsecant of the slope angle by two. The exsecant is found by formula 8 on p. 218 *Standard Field Tables* (SFT). The table is usable for feet or any other unit. For 100 ft. the extra distance is *half* the table figure. Figure odd distances by interpolation.

For stations 2 ch. horizontal, a topographic trailer tape and a topographic clinometer are needed. The tape has an extension on the rear with the points for various slopes for 1 ch. marked on one side and 2 ch. on the other. The compassman allows the tape to be pulled beyond the normal 2-ch. point to the mark corresponding to the slope reading. Topographic trailer tapes have several disadvantages: they are more expensive, require more time to use efficiently, to handle and "throw", and tend to be less accurate.

Obstacles on the Line. A tree or other obstacle will often be in the line of sight. If it is a rock or brow of a hill or down log, the chainman should usually drop back to where he can be seen. If the rod is invisible because of a tree or brush, several solutions are available. Waving the rod slowly is often good. If the surveyor can see the rod by moving his head a foot or so to either or both sides of the compass, he can direct the chainman by eye to reset the rod. If the rodman is visible but the rod is not, use the center of the rodman or hold the rod horizontally so equal parts of the rod are visible on each side of the obstacle. Another method is to reset the compass a few feet beyond the obstruction. If a 2″ hollow aluminum range pole is used, carry in it a 27″ metal venetian blind slat marked at the midpoint. If the pole ends up behind a tree, the rodman can remove the slat and hold it horizontally with half on each side of the pole. The compassman estimates the pole position by sighting on the slat.

As these methods become more time-consuming, move the compass 1, 2, or 3 links to one side or the other at a right angle determined by eye. The forward measurement is made as usual and a stake set on the offset line, then offset back to the original line. The offset is not mentioned in the notes.

Fieldbook Record. The left hand page should look like this:

RESURVEY OF DEED OF 1/16/42 (182/433 Rand. Co.) for Harry R. Jones. Beg. at a 22″ red oak said by Jones to be 4th cor. of deed, about 100 ft. N of main Fks. of Files Cr.

STA.	*MAG. BEAR.*	*SLOPE DIST.*	*VERT. ANGLE*	*HOR. DIST.*	*OFFSET*
RO				ch.*	
	N 30 W	0.25	0		
0					
	N 60 E	2.000	+4½	1.993	
2					
	N 60 E	2.000	0	2.000	

Note: Allow an extra space between entries. *Figured later.

Subsequent Lines. The compass is set over the new station, levelled, and sighted on the backsight. The chainman passes the tape around an obstacle to keep it from the tripod, and moves ahead on the probable line of sight, looking back over the line just run to orient himself. A good chainman will go a chain or so before he needs help from the compass. The surveyor makes final adjustment of the compass, then writes the station number on the flat part of the stake, ties flagging on it and sees that it is vertical. Each station is numbered to correspond to its slope distance from the beginning. If the terrain does not allow a 2-ch. shot, shorten the distance to 1 or 1½ ch.; stakes are numbered accordingly. From here on, each shot is a repetition of the above.

Backsight Control. If the foresight bearing is not in line with the backsight, it is possible (1) the backsight stake may be tilted, or (2) the needle is sticking, or (3) the compass may have been set a little to one side on one station, or (4) the compass was not level. If the difference is more than 5 m. and cannot be corrected, the backsight should usually control, and a note is made of the back and foresight bearings. This difference should last for only one shot and might be caused by local attraction. This type of trouble deserves the most careful handling. (See chapter on magnetic variation.)

Right Hand Page Notes. Record on the right hand page, across from the proper stations, the distance at which the center of streams, fences, roads, etc., are crossed. If a station point is within 20 ft. of a wire fence, the line distance should be shortened or lengthened to avoid attraction, or the backsight used; otherwise an offset is required.

The Last Shot. Just before the last shot the surveyor should reduce all the slope distances to horizontal (unless deed calls for surface measurement), with the accompanying tables (or traverse tables):

SLOPE REDUCTION TABLE

DEGREE	2.000	1.750	1.500	1.250	1.000	DEGREE
3	1.997	1.748	1.498	1.248	0.999	3
½	96	47	97	48	98	½
4	1.995	1.746	1.496	1.247	0.998	4
½	94	45	95	46	97	½
5	1.992	1.743	1.494	1.245	0.966	5
½	91	42	93	44	95	½
6	1.989	1.740	1.492	1.243	0.995	6
½	87	39	90	42	94	½
7	1.985	1.737	1.489	1.241	0.993	7
½	83	35	87	39	91	½
8	1.981	1.733	1.485	1.238	0.990	8
½	78	31	84	36	89	½
9	1.975	1.728	1.482	1.234	0.988	9
½	73	26	79	33	86	½
10	1.970	1.723	1.477	1.231	0.985	10
½	67	21	75	29	83	½
11	1.963	1.718	1.472	1.227	0.982	11
½	60	15	70	25	80	½
12	1.956	1.712	1.467	1.223	0.978	12
½	53	08	64	20	76	½
13	1.949	1.705	1.462	1.218	0.974	13
½	45	02	59	16	72	½
14	1.941	1.698	1.455	1.213	0.970	14
½	36	94	52	10	68	½
15	1.932	1.690	1.449	1.207	0.966	15
½	27	86	45	04	64	½
16	1.923	1.682	1.442	1.202	0.961	16
½	18	78	38	198	59	½
17	1.913	1.674	1.434	1.195	0.956	17
½	07	69	31	92	54	½
18	1.902	1.664	1.427	1.189	0.951	18
½	897	60	22	85	48	½
19	1.891	1.655	1.418	1.182	0.946	19
½	85	50	14	78	43	½
20	1.879	1.644	1.410	1.175	0.940	20
½	73	39	05	71	37	½
21	1.867	1.634	1.400	1.167	0.934	21
½	61	28	396	63	30	½

SLOPE REDUCTION TABLE

DEGREE	2.000	1.750	1.500	1.250	1.000	DEGREE
22	1.854	1.623	1.391	1.159	0.927	22
½	48	17	86	55	24	½
23	1.841	1.611	1.381	1.151	0.921	23
½	34	05	76	46	17	½
24	1.827	1.599	1.371	1.142	0.914	24
½	20	92	65	38	10	½
25	1.813	1.586	1.359	1.133	0.906	25
½	05	80	54	28	03	½
26	1.798	1.573	1.348	1.124	0.899	26
½	90	66	42	19	95	½
27	1.782	1.559	1.337	1.114	0.891	27
½	74	52	31	09	87	½
28	1.766	1.545	1.324	1.104	0.883	28
½	58	38	18	098	79	½
29	1.749	1.531	1.312	1.093	0.875	29
½	41	23	06	88	70	½
30	1.732	1.516	1.299	1.082	0.866	30
½	23	08	92	77	62	½
31	1.714	1.500	1.286	1.072	0.857	31
½	05	492	79	66	53	½
32	1.696	1.484	1.272	1.060	0.848	32
½	87	76	65	54	43	½
33	1.677	1.468	1.258	1.048	0.839	33
½	68	59	51	42	34	½
34	1.658	1.451	1.244	1.036	0.829	34
½	48	42	36	30	24	½
35	1.638	1.434	1.229	1.024	0.819	35
½	28	25	21	18	14	½
36	1.618	1.416	1.214	1.011	0.809	36
½	08	07	06	05	04	½
37	1.597	1.398	1.198	0.998	0.799	37
½	87	88	90	92	93	½
38	1.576	1.379	1.182	0.985	0.788	38
½	65	70	74	78	83	½
39	1.554	1.360	1.166	0.971	0.777	39
½	43	50	57	64	72	½
40	1.532	1.340	1.149	0.958	0.776	40
½	21	31	41	50	60	½

When the Hor. Dist. is recorded for all the shots, they are added, then subtracted from the planned length of the line. The remainder is the Hor. Dist. of the last shot. The surveyor allows the chainman to proceed to that Dist. and set a temporary stake. Record this in the proper column. If this Dist. was measured on the slope, the extra Dist. is reduced to Hor. Dist. and the stake moved accordingly. The last stake is now at a bearing and Hor. Dist. called for in his calculation.

The surveyor now searches for corner evidence. If the corner is found, the random line is extended or shortened to a point at right angle to the corner and the length of the last shot adjusted accordingly. Do not erase. The right angle point can be found by setting the compass over the corner or by trial and error movement on the random line. The bearing and distance of the offset to the new corner is measured to the nearest tenth link and recorded. Vertical angle of the offset is recorded if the measurement is slope. The compass is then set over the new corner and the next line may be run.

Surveying With 3 or 4-man Crew

Here the duties are divided differently. A third man usually cuts and shapes the stakes and places the number and flag thereon. The third man helps cut brush. A fourth man cuts brush and may be used as a backsight man. As the size of the crew increases, the output per man drops. If 2 men run 10 ch. per hr., 3 will not run 15 nor will 4 run 20.

Plotting the Random Survey

Random lines may be plotted on several scales. Major importance should be given to a scale large enough to show the error of closure within reasonable limits.

The closing call will seldom return exactly to the beginning, even though the field survey did return thereto. The distance between the beginning and end of the last line is the E.C. To see if the E.C. falls within the limit of accuracy set for the survey, add the length of all lines in the random survey, including offsets, and divide this total by the distance by which the plat did not close. Example: 235 ch. were run and the plat lacked 0.15 ch. of closing. 235 divided by 0.15 = 1566 or an error of 1 in 1566. If the E.C. is bad, a field check may be needed.

At this point the surveyor must decide where lost and obliterated corners are to be located. A corner is *obliterated* if the original monument is missing but other evidence is available to locate it; a corner is *lost* if *no* evidence is available to locate it. It is here that skill, knowledge, and experience are brought to bear on the problems. The weight of all evidence is considered in the light of the law and court decisions. When the decisions are made, the surveyor is ready to calculate the true lines leading to final action on the ground.

Chapter 23

THE TRUE LINES

Calculating the True Line

When Two Adjacent Corners Are Found. If the corners are known at each end of the line, and the random line has missed the second corner (after offsets are allowed for), the bearing and length of the true line between the corners is now calculated. (This "true" has no connection with "true north".) Although the true line could be scaled from a large-scale plat, it is better to use the formula at p. 218 SFT (p. 23.14 *Forestry Handbook*). On the diagram at the left, (A) = the beginning corner, (B) the true corner, (C) the end of the random line (b), (a) the offset to the true corner, and (c) the length of the true line. Record this triangle and calculations in the field notes. The angle at (A) is found by solving this formula:

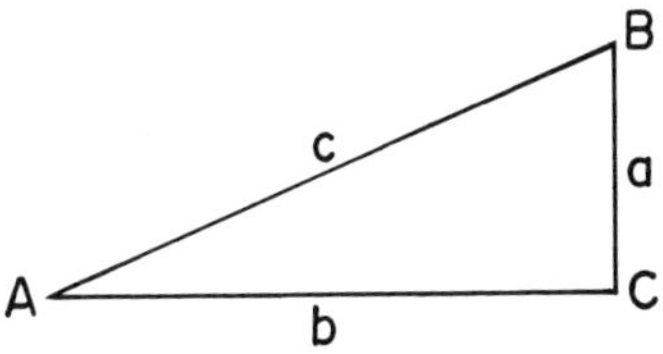

$$\text{tangent of A} = \frac{a}{b}$$

Assume a random line of 41.33 ch. and an offset of 0.144 ch. Then .144 divided by 41.33 = .003484. In the table of natural tangents (p. 115 SFT) .00348 is just a little under 12′. Angle A can be found in another way: convert the offset to links per 10 ch. and look up the result in the table on p. 7 of SFT. Thus 14.4 divided by 4.133 = 3.48 which = 12′. If the random line was run on N 60 E and ended to the right or SE of the true corner, the bearing of the true line (c) = N 60 E minus 12′ or N 59-48 E.

If there was a preliminary offset before starting the random line, this offset is subtracted from or added to the offset at the second corner. Whether to add or subtract is best told by a small diagram in the notes. Show all offsets made on the line. The following diagram shows 3 offsets in one direction totaling 55.6 and one in the other direction of 70 l. (all horizontal, of course), leaving a new offset of 14.4 links.

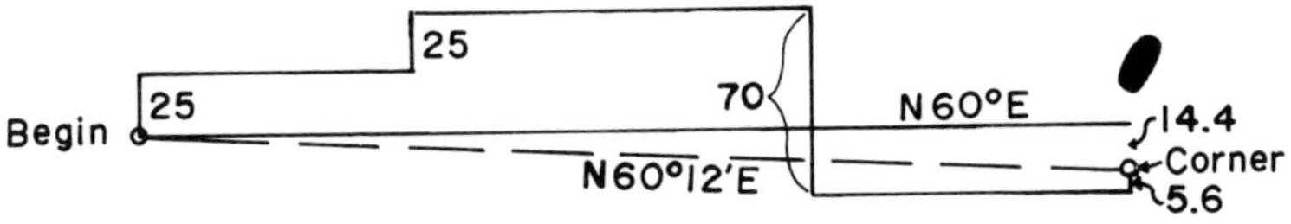

If north be up, the angle represented by 14.4 l. and the length of the random line would be turned clockwise, the true line is N 60-12 E.

The length of true line (c) can be found by either of 2 formulas (p. 218 SFT). First: c = a divided by sin A; or .144 by .00349 = 41.26. Second: b divided by cos A; or 41.33 by .99999 = 41.33. It is obvious that something is wrong because the hypotenuse of a right triangle *must be longer* than any other side. The error is this: the tangent of (A) is actually a little less than 12′. The sine of the real angle is .003484, which divided into .144 = 41.332.

The significance of the angle is this: succeeding random lines may come closer to the succeeding corners if the bearings of the random lines are adjusted to the right or left, depending on the direction of miss of the first line.

When Two Consecutive Corners Not Found. If the second corner cannot be found, no offset or corrective angle can yet be calculated for that line. In that case the second random line begins where the first one ended. Authentic line marks or line fences near the random line may affect the location depending on the law and the facts. The distance to such marks is used to calculate corrective angles as in the example above.

If there is no evidence that more weight should be given to one line than another, lost corners can be determined by single or double proportionate measurement. This is usually the method of last resort. This strictly mathematical relocation of a lost corner is not based on any legal decision, but is obtained by allowing every known corner within reasonable radius to enter into its control, each corner being given a weight inversely proportional to its distance from the missing corner. The single proportionate method is described on p. 267 of TM5-235 and in Chap. V of the 1947 *Manual*. The theory is to measure the bearing and distance from the end of the last random line to the authentic corner, and to distribute that distance among the random lines on the basis of the proportion which each line bears to the total length. Each new station is located on the same bearing as the bearing of the closing tie line to the authentic corner. Below is an example:

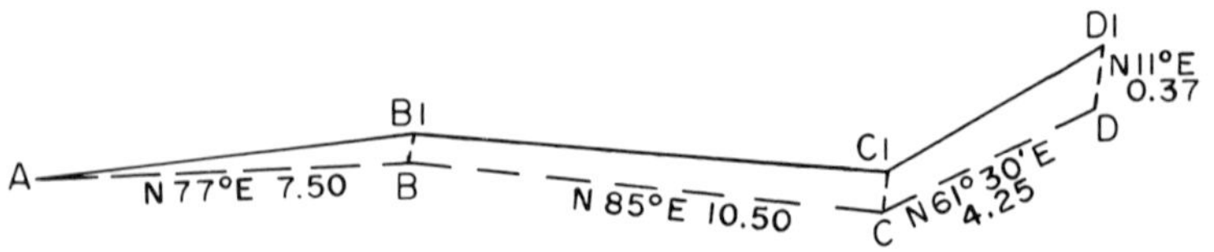

The first 2 random lines are run on the calculated bearing and deed call distance, shown by dash lines. They fail to disclose any corners, but at the end of the third line, a tie is made to a known corner (D1). The total distance run is 22.25 ch. The 37 l. are distributed among the 3 lines thus: for every chain run, the distribution would be .37 divided by 22.25, or .0166 ch. Since the first line was 7.50 ch., that line would be given 7.50 x 1.66 or 12.45 links; this distance is to be measured from (B) on a bearing of N 11 E. The second line gets 29.88 l. which is measured N 11 E from (C). Stakes are set at (B1) and (C1) and monumented as corners. If some evidence of the true corner was found in the vicinity, that, of course, would control over measurements. The proportionate system is merely prorating the EC among a series of lines—just a variation of the principle of interpolation.

Either now or later the bearing and distance must be computed for the true lines above (A-B1-C1-D1). These can be found for (A-B1) in several ways.

1. The triangle below can be plotted with drafting machine on a scale large enough to scale the answer from the plat. The scale should allow the error to be not over 2 or 3 minutes or 1 link of distance. In this example, 5″ to 1 ch. would make a plat about 40″ long. If paper of this size is not handy, two smaller pieces of paper can be fastened to the drafting table about 35″ apart. Point (A) is placed on the left sheet so that points (B) and (B1) will fall on the other sheet.

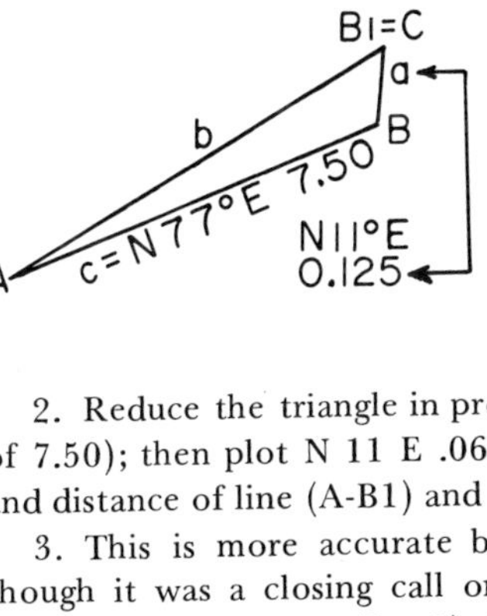

2. Reduce the triangle in proportion. Example: plot N 77 E 3.75 (½ of 7.50); then plot N 11 E .0625 (½ of 12.5 l.); now scale the bearing and distance of line (A-B1) and double it.

3. This is more accurate but takes more time. Figure side (b) as though it was a closing call or E.C. (see Sec. 169 of TM5-232 or p. 17.24 of *Forestry Handbook*). Use a calculator. The example works out as follows:

			Latitude	*Departure*			
				cos x dist.		*sin x dist.*	
Bear.	*Dist.*	*Cos*	*Sin*	*N*	*S*	*E*	*W*
N 77 E	7.50	.22495	.97437	1.687		7.308	
N 11 E	.125	.98163	.19081	.123		.023	
				1.810		7.331	

Tangent of the missing bearing is the difference between the E and W departures divided by the difference between the N and S latitudes. Thus 7.331 divided by 1.810 = 4.05027, which is the tangent of 76-08 (or a little less) p. 118 SFT. Bearing of line (A-B1) then is N 76-08 E. Length of line (A-B1) equals the Dep. divided by sin of the bearing, or 7.331 by .97086 = 7.55 ch. (p. 108 SFT).

4. This requires still more time with no increase in accuracy. Use the formulas for solving an oblique triangle (p. 219 SFT). Angle (B) in diagram under (1) above can be figured from the bearings: 90 - 77 + 90 + 11 = 114°. But with only (B), (a), and (c) known, none of the formulas on p. 219 SFT are usable. The letters on the triangle must be rearranged so one of the *given* combinations on p. 219 is usable. The best arrangement is the one at the left.

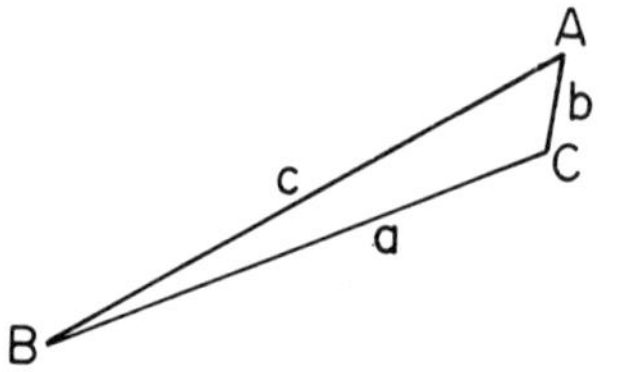

Then: ½(A+B) = 90 - ½C, or 90 - ½ of 114, or 90 - 57 = 33. Also: tan ½(A - B) =

$$\frac{a-b}{a+b} \text{ x tan } ½(A + B), \text{ or } \frac{7.50 - .125}{7.50 + .125} \text{ x tan } 33°, \frac{7.375}{7.625} \text{ x } .64941, \text{ or}$$

.96721 x .64941 = .62812, which is the tangent of 32-08.

We now have the data to solve for angles (A) and (B).
(B) is ½(A + B) - ½(A - B), or 33 - 32-08, or 52′.
(A) is ½(A + B) + ½(A - B), or 33 + 32-08, or 65-08.

To check, the 3 angles must add up to 180 degrees. Bearing of line (c) (originally b) is therefore N 77 E minus 00-52 or N 76-08 E. Now to obtain the length of line (c):

$$c = (a + b) \text{ x } \frac{\cos ½(A + B)}{\cos ½(A - B)} \text{ or } 7.625 \text{ x } \frac{\cos 33}{\cos 32\text{-}08}$$

$$\text{or } 7.625 \text{ x } \frac{.83867}{.84681} \text{ or } 7.625 \text{ x } .99039 = 7.552$$

There is a different situation in regard to the bearing and length of line (B1-C1), because the figure is 4-sided instead of a triangle. The diagram left below shows this problem. Draw a line from (B) to (C1). Angle (C-B-C1) is solved exactly as for triangle (A-B-B1), this gives the bearing of line (B-C1). Length of (B-C1) is found the same way as for (c) above. We now have triangle (B-B1-C1) with two sides known. Angle (B1-B-C1) is found by subtract-

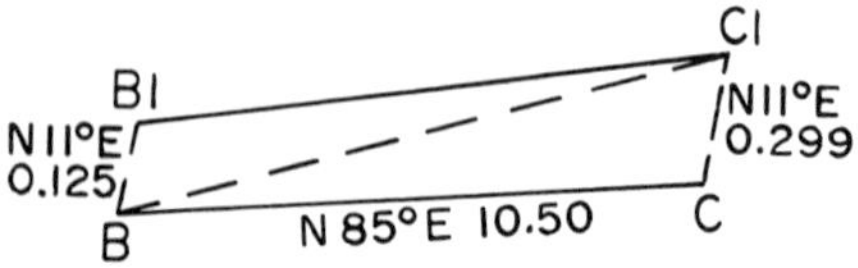

ing the bearing of line (B-B1) from the bearing of line (B-C1). Length and bearing of (B1-C1) is found the same way as for (A-B1). Length and bearing of (C1-D1) (p. 32) is found the same way.

If the formula for an oblique triangle calls for the sine of an angle greater than 90°, substitute the cosine of the angle minus 90. If the angle is 114 and (b) is sought, the formula (b) = a divided by sin A x sin B would become a divided by sin A x cos 24°.

All work sheets of formulas and calculations should be clearly identified in red pencil and filed. Note: For small lots, less time may be needed to measure lines on the ground than to calculate them as oblique triangles.

Correcting Errors. The sum of the interior angles should equal 90 x twice the number of angles minus 4. If the EC in a random survey is worse than 1 in 1000, the surveyor should check his field methods.

Locating the True Lines

After deciding the location, bearing and length of the true lines, how are they established on the ground? Prorate the number of links (or feet) by which the random line missed the corner, over the several shots required to reach the corner. Example: if a line 1000 feet long missed a corner by 10 ft., the 500 ft. point on said line would miss the true line by 5 ft. The same principle (interpolation) is used to distribute the offsets among 2-ch. or any length shots.

Example: A 41.33 ch. line missed the corner by 0.144 ch. Convert this to feet: 0.144 x 66 = 9.504′. 9.50 is then placed in the offset column in the field notes opposite the 41.33 ch. station. Amount of offset per chain is obtained by dividing 9.504 by 41.33; result 0.23 ft. If the distance from station zero to station 2 was 2 ch. Hor., the offset would be 2 x 0.23 or 0.46 ft. at station 2. If the Hor. Dist. from station 2 to 4 was 1.981 ch., the offset would be 1.981 x 0.23 = 0.46 ft. plus 0.46 from the previous shot, or 0.92 ft. offset at station 4. This type of calculation is repeated for each shot by setting the constant figure of 23 in the calculator and multiplying it by each Hor. Dist. The offset need not be recorded closer than 0.01 ft. or even 0.1 ft.

When all offsets are entered, a cumulative total is run on the calculator and entered in the notes like this:

STA.	*Mag. Bear.*	*Slope Dist.*	*Vert. Angle*	*Hor. Dist.*	*Offset feet to NW*
0					
	N 60 E	2.000	0	2.000	0.46
2					0.46
	”	2.000	+8	1.981	0.46
4					0.92
	”	1.500	+4-½	1.495	0.34
5.5					1.26

Direction of offsets should be recorded at the head of the column; if direction should change because of obstacle offsets in the random line, note the change at the proper point. Converting offsets to feet and fractions thereof allows offsets to be measured with a pocket tape instead of dragging a 2-ch. tape. When offsets are added, the final figure is compared with the figure already entered in the notes as the last offset. If there is a difference of more than ½ ft. find the source of the error; something may be wrong with the procedure.

The random stakes should now be offset to the true line. It is easier to start at the zero end, but the starting point will usually be determined by the overall plan of operation.

Small Offsets. Small offsets can be measured by one man. He sticks a pin or stick through the tape loop and then into the ground so zero will be at the random stake, then pulls the stake and sets it over the amount of the offset. The pin is then pulled up and tape rolled in, ready for the next offset. If the offsets are on level ground or downhill, one man can handle it easily, "breaking tape" if needed. If the offset is uphill, estimate the offset and stick the pin and tape on the uphill end, bring the tape over the random stake by eye and read the tape; measure the difference between the estimated offset and the true offset, either up or downhill, and move the stake to the right point. A large spike may be set at the first 2 or 3 stakes.

Longer offsets.

1. If the land is level or nearly so, the surveyor stands with one toe at the random station mark and sights about a right angle with a hand compass, selecting some point at least as far away as the length of the offset. An assistant takes the stake and one end of the tape, proceeds to the offset distance on the proper line of sight. The tape is held level by eye over whichever point is lower and the stake is set.

2. If the land is steep, the slope distance can be calculated and measured on the slope or several short Hor. lengths can be measured.

3. When offsets become too long, too steep or too rough, it may be better to run a new random line on the calculated bearing of the true line. This new line will likely hit within a short distance of the corner, and offsets are easy.

When offsetting, look for old marks of previous surveys and for possession; they carry weight in controlling lines.

The Monongahela National Forest computes offsets on a transit survey as shown on the following page.

COMPUTATION OF OFFSETS FROM RANDOM TRAVERSE LINE TO TRUE LINE

Tract Name Maryland and West Virginia Tract No. 99-P

Corner 60 to Corner 61

True Bearing N 30°38'12" E 22.966 Chains

Computed by: G.A.B. Checked by: B.R.R. Date 9-5-63 Sheet 1 of 1

STATION	BEARING	DISTANCE	∠ DIFFERENCE	SINE	OFFSET	TOTAL OFFSET
Cor. 60						
	N 43° 27' E	0.291	12°48'48"	0.22177	0.065 L	
M-45						0.065 L
	N 33° 45' E	1.333	03°06'48"	0.05431	0.072 L	
M-46						0.137 L
	N 33° 45' E	1.312	03°06'48"	0.05431	0.071 L	
M-47						0.208 L
	N 33° 45' E	2.235	03°06'48"	0.05431	0.121 L	
M-48						0.329 L
	N 33° 45' E	1.121	03°06'48"	0.05431	0.061 L	
M-49						0.390 L
	N 33° 45' E	0.827	03°06'48"	0.05431	0.045 L	
M-50						0.435 L
	N 33° 45' E	0.872	03°06'48"	0.05431	0.047 L	
M-51						0.482 L
	N 33° 45' E	2.049	03°06'48"	0.05431	0.111 L	
M-52						0.593 L
	N 33° 45' E	0.918	03°06'48"	0.05431	0.050 L	
M-53						0.643 L
	N 33° 45' E	3.804	03°06'48"	0.05431	0.207 L	
M-54						0.850 L
	N 33° 45' E	4.069	03°06'48"	0.05431	0.221 L	
M-55						1.071 L
	N 16° 40' E	4.317	13°58'12"	0.24142	1.042 R	
M-56						0.029 L
	S 86° 15' W	0.035	124°23'12"	0.82525	0.029 L	
Cor. 61						

Chapter 24

LATITUDES AND DEPARTURES

Latitude is the distance which one end of a line is N or S of the other end of the same line. Departure is the distance which one end of the line is E or W of the other end. These distances are sometimes called "coordinates". Lats. are plus when the bearing is N and minus when the bearing is S. Deps. are plus for E bearings and minus for W bearings. To explain this, plot, as at the left, at 10 units per inch, line AB N 55 E 1.00″ long. Then (B) is about 0.57″ N of A when measured from (C). The Lat. of N 55 E is .57 x length; if the line was 2 ch. long, the Lat. is 1.14 ch. N. (B) is about 0.82″ from (A) when measured from (D). Thus the Dep. is .82 E. As the bearing of a line decreases, the Lat. increases and the Dep. decreases.

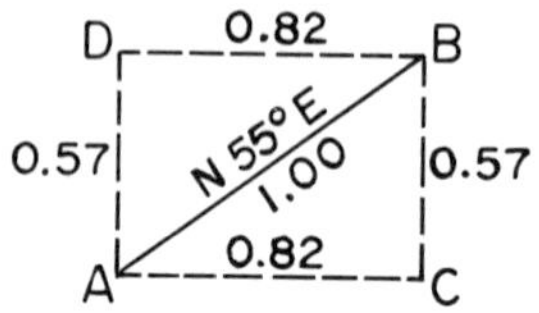

Tables (p. 106 SFT) have been made to show what is the Lat. and Dep. for a line with a length of one unit, for each degree and minute from a N-S line up to 90°. The cos of the bearing x the length of the line equals its Lat.; the sin x length equals its Dep. Diagram and formulas shown on p. 218 SFT. If distance is measured to the nearest link or foot, it is useless to carry out the Lats. and Deps. beyond a tenth of a link or foot. Example: a 1.00 chain line at N 55 E has a Lat. of .57358. But we need not use more than .574 because .0005 ch. is less than ½″.

If the Lat. and Dep. can be figured for each line, and if several lines are joined end to end as in a survey, the Lats. and Deps. for each line can be added and give a cumulative figure (coordinates) for any given point. If a survey is a closed traverse, the N and S Lats. should be equal, also the E and W Deps.

The Lats. and Deps. for a closed traverse will seldom balance out to zero because of inaccuracies of field work. The difference between the N and S Lats. and between the E and W Deps. will show the EC thus: Square each difference, add them together and take the square root. This EC is expressed as 1 in ———, by dividing the sum of the length of all lines run, by the EC (see following chapter).

A *traverse table* is merely a table of natural sines and cosines for distances up to 100; table 4 in SFT is figured for ¼ degree. For uneven distances, break them down and adjust decimal points. Whether traverse tables or the sin-cos method is used depends on which is easier and gives the accuracy desired.

Chapter 25

FINAL CALCULATIONS

Error of Closure

Before making the final plat and description of survey, the EC and area must be determined. Two methods can be used:

1. A plat of the true calls is made on a scale large enough to measure the EC with a ruler.

2. Calculate the EC by Lats. and Deps. This method is necessary for surveys requiring greater accuracy. Assume a survey with 5 calls below. Arrange the calls so the first call begins at the most westerly point of the survey. Some surveyors use a special form with headings already printed on them.

The cos and sin of each bearing is inserted in the proper column. Multiply with machine the distances by the cos and sin and place answers in the proper columns. Round off to 3 places. Now add the four Lat. and Dep. columns. Subtract the smaller Lat. and Dep. from the larger one. The Lats. fail to close by .072 and the Deps. by .218.

				Latitude		*Departure*	
	Poles	*Cos of*	*Sin of*	*cos x dist.*		*sin x dist.*	
Bearing	*Dist.*	*bearing*	*bearing*	*N*	*S*	*E*	*W*
N 27-45 E	12.79	.88499	.46561	11.319		5.955	
N 76-08 E	30.20	.23966	.97086	7.238		29.320	
S 46-30 E	23.62	.68835	.72537		16.259	17.133	
S 51-30 W	25.40	.62251	.78261		15.812		19.878
N 67-41 W	35.40	.37973	.92510	13.442			32.748
	127.41			31.999	32.071	52.408	52.626
					-31.999		-52.408
					.072		.218

The square root of .072 squared + .218 squared, is the square root of .005184 + .047524, or the square root of .052708 = .230, or an error of 1 in 554.

Determination of Area

Balancing the Latitudes and Departures. Balance the Lats. and Deps. so northings equal southings. This can be done in two ways:

Compass Rule.

This is the most common rule. The total distance 127.41; the result is the error in one pole: .0005651. Multiply .0005651 by each distance. The result is added to each N Lat. and subtracted from each S Lat. so the sum of the norths will equal the sum of the souths. Put the figures in the Balanced Lat. column with the proper sign in front (see below). Check the result by adding the northings (32.043) and comparing with the sum of the southing (32.044). The difference of .001 is due to fractions and is added to or subtracted from the figure that has the most variation. Do the same for the Deps.

Balanced by Compass Rule		*Balanced by Transit Rule*	
Lat.	*Dep.*	*Lat.*	*Dep.*
+11.326	+ 5.977	+11.332	+ 5.967
+ 7.255	+29.372	+ 7.246	+29.381
-16.246	+17.173	-16.241	+17.168
-15.798	-19.835	-15.794	-19.837
+13.462	-32.688	+13.457	-32.680
+32.043	+52.522	+32.035	+52.516
-32.044	-52.523	-32.035	-52.517

Transit Rule.

Divide .072 by the sum of north and south Lats. (64.070) to get the error in one unit of Lat. (.0011238). Multiply .0011238 x each Lat. and add or subtract the result from the proper Lat. so as to make the norths equal the souths. Example: .0011238 x 11.319 = .013, which + 11.319 = 11.332. Do the same for the Deps. The procedure is generally similar to the first method, including the final adjustment.

The question arises: Should bearings be corrected to correspond to the balanced Lats. and Deps.? The answer is generally "no". Example: correct bearing N 27-45 E by working the calculation backward. Divide Lat. 11.326 by 12.79; the result .88554 is the cos of the sin of 27-52. An average is 27-46.5, which is only 1½′ off from 27-45. Unadjusted bearings allow a check of the former surveyor's precision.

Use of 4-place Tables. A 4-place table shows 27-45 to have a cos of .8850, which x 12.79 = 11.319—identical with the 5-place result. Even if the difference is as much as .004 in cos of 76-08, .2397 x 30.20 = 7.239, which is almost the same as 7.238.

Use of 3-place Tables. Here .88499 = .885, which x 12.79 = 11.319. However, .23966 becomes .240, which x 30.20 = 7.248. This is only .01 pole off from 7.238, or about 2″.

Rounding Off.

The rule (p. 374 of *Tracy*) is: "for 2 certain figures in a result, use 3-place tables; for 3 certain figures, use 4-place tables" and so on. If the last number to be eliminated is a 5, add 1 to the last figure to be kept if this will make it even; otherwise leave it unchanged.

Calculation for Area

Four methods to determine area are: by right triangles, planimeter, double meridian distance and coordinates.

Right Triangles. Divide the area into right triangles.

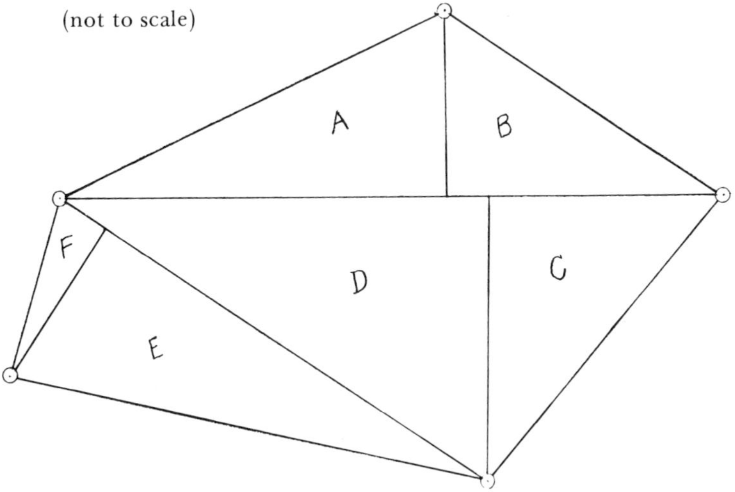

With a fine scale rule, say 50 per inch, measure the 2 short sides of each triangle, multiply them, add and divide by 2.

A	138	x	63	=	8694.0
B	98½	x	63	=	6205.5
C	98	x	82½	=	8085.0
D	98	x	154	=	15092.0
E	62	x	164	=	10168.0
F	62	x	18	=	1116.0
					49360.5 divided by 2 is 24680.

24680 is the number of square units in the plat with the 50 scale, which has 2500 units per square inch. Then 24680 divided by 2500 = 9.872 sq. in. Since the map scale is 10 p. per in., a sq. in. is 100 square poles. Then 9,872 sq. in. = 987.2 sq. poles. As there are 160 sq. poles in an acre, 987.2 divided by 160 = 6.17 acres. This method can be shortened by multiplying the common altitude of adjoining triangles, by the combined lengths of their bases (for A and B, 63 x 138 + 98½).

Planimeter. If the machine runs off the paper in a practice run, put a piece of tape on the paper edge so the roller of the planimeter will run smoothly back on the sheet. Set the roller at zero and the pointer on a corner and go clockwise around the survey at least 3 times. If the readings are 9.85, 19.72, and 29.55, the average is 9.85 (none being far enough off to reject). Then 9.85 sq. in. = 985 sq. p. divided by 160 = 6.156 acres. Be sure the planimeter wheel does not slip on the paper.

Double Meridian Distance. An example using the previous Compass Rule figures:

Balanced latitude	*Balanced departure*	*DMD*	*Double Area* +	–
+ 11.326	+ 5.977	+ 5.977	67.6955	
+ 7.255	+ 29.372	+ 41.326	299.8201	
– 16.245	+ 17.173	+ 87.871		1427.4644
– 15.798	– 19.835	+ 85.209		1346.1318
+ 13.462	– 32.687	+ 32.687	440.0324	
			807.5480	2773.5962
				– 807.5480
				1966.0482

Then 1966.0482 divided by 2 = 983.0241 square poles, divided by 160 = 6.144 acres.

The figures in the DMD column are calculated by these rules:

The DMD of the first course is its departure
The DMDs of later courses are:

a. The DMD of the preceding course plus
b. The departure of the preceding course plus
c. The departure of the course itself

Thus the DMD of the second course is 5.977 + 5.977 + 29.372 or 41.326. The third course is 41.326 + 29.372 + 17.173 or 87.871. And so on. If the last DMD equals the last Dep. and is of the opposite sign, the DMD calculation is correct.

Calculation of Double Area. Multiply the DMD of each course by the Lat. of that course, observing both algebraic signs (+ x + = +; - x + = -; - x - = +). For the first course, 5.977 x 11.326 = 67.6955 which is placed in the + column. So on for the other courses. Now total the pluses, total the minuses, subtract the smaller from the larger and divide by 2. If the survey was in poles, the result is square poles, or 6.144 acres. The more care used in the first two methods, the closer their results will be to the DMD result.

Coordinates

Some surveyors calculate coordinates for their surveys. Coordinates can be used to:

1. Plot a traverse (instead of using a ruler and protractor or drafting machine).
2. Check accuracy of balanced latitudes and departures.
3. Tell how far to offset a random stake to the true line.
4. Determine area (perhaps easier than by DMDs.

Coordinates express the location of corners of a traverse with reference to a given point to the corners. For example, if the beginning corner of the previous traverse is assumed to be zero, then corner 2, at the end of the N 27-45 E 12.79 call, would have an X coordinate equal to the latitude, or 11.319; and a Y coordinate equal to the departure, or 5.955 (when balanced 11.326 and 5.977). These are plus values because they are north and east of the zero point. The X coordinate for the third corner would be 11.326 plus 7.255, or 18.581; the Y coordinate would be 5.977 plus 29.372, or 35.349.

The plus values eventually change to minus values, but these are cumulative and will end up as zero if calculations are correct. An assumed coordinate large enough to overcome minus values is sometimes used for the beginning corner. The computation of coordinates for the traverse is made as follows, assuming corner 1 to have a coordinate of zero, and all figures being plus unless minus is shown:

Corner	*X (Lat)*	*Y (Dep)*
1	0	0
	11.326	5.977
2	11.326	5.977
	7.255	29.372
3	18.581	35.349
	-16.245	17.173
4	2.336	52.522
	-15.798	-19.835
5	-13.462	32.687
	13.462	-32.687
1	0	0

Conversion of these coordinates into area is described below in two ways:

1. Each X coordinate is multiplied by the difference between the two *adjacent* Y coordinates. The sum of each multiplication is added, and the total equals twice the area in the same unit used in measurement of distance. This is expressed by formula thus:

Double area equals X1 x (Y5-Y2) + X2 x (Y1-Y3) + X3 x (Y2-Y4) + X4 x (Y3-Y5) + X5 x (Y4-Y1). And so on, if there are more sides. Here is the calculation for the above coordinates:

Corner 1: X coordinate is 0, which x (32.687-5.977)	is	0
Corner 2: 11.326 x (0-35.349) or 11.326 x (-35.349)	is -	400.336

Corner 3: 18.581 x (5.977-52.522) or 18.581 x (-46.545)	is - 864.853
Corner 4: 2.336 x (35.349-32.687) or 2.336 x 2.662	is 6.218
Corner 5: -13.462 x (52.522 - 0) or -13.462 x 52.522	is - 707.051
	Total is -1972.267 + 6.218 or -1966.049

Drop the minus sign. Divide 1966.049 by 2 = 983.025 square poles, divided by 160 (square poles in one acre) equals 6.144 acres.

This method may be simplified by rounding off the Lats. and Deps. to two places, if three is too exact for the techniques involved:

Corner	*X (Lat)*	*Y (Dep)*
1	0	0
	11.33	5.97
2	11.33	5.97
	7.25	29.38
3	18.58	35.35
	-16.24	17.17
4	2.34	52.52
	-15.80	-19.84
5	-13.46	32.68
	13.46	-32.68
1	0	0

Then:

Corner 1: 0 x (32.68-5.97)is	0
Corner 2: 11.33 x (0-35.35)is -	400.516
Corner 3: 18.58 x (5.97-52.52)is -	864.899
Corner 4: 2.34 x (35.35-32.68)is	6.248
Corner 5: -13.46 x (52.52-0)is -	706.919
Total is	-1966.086
Divided by 2 is	983.043
Divided by 160 is	6.144

2. If a hand calculator like the Monroe is used, the job can be simplified by leaving cumulative figures in the machine as follows:

Corner 1: Zero is the answer, so pass on to corner 2.

Corner 2: Set up 11.33 in machine; multiply forward by 0-35.35 or -35.35; machine reads 400.5155
Leave this in machine and remove 11.33 & 35.35

Corner 3: Set up 18.58 in machine; multiply forward by 5.97-52.52 or -46.55; machine reads 1265.4145
Leave in machine and remove 18.58 and 46.55

Corner 4: Set up 2.34 in machine; multiply *backward* by 35.35-32.68 or 2.67; machine reads 1259.1667
Leave in machine and remove 2.34 and 2.67

Corner 5: Set up 13.46 in machine; multiply forward by 52.52-0 or 52.52; machine reads 1966.0859

which divided by 2 is 983.043, divided by 160 equals 6.144 acres.

Note: Turn handle of calculator forward when the Lat. x Dep. is minus, and backwards when the product is plus.

Another formula as shown on p. 17.25 of the *Forestry Handbook* is to add the Lat. of the corner to the Lat. of the following corner, and multiply by the Dep. of the course. Let the totals for each operation accumulate in the machine. Here is how it works:

			Machine reads
0 + 11.33 x 5.97 is	67.6401		67.6401
11.33 + 18.58 x 29.38 is	878.7558 +	67.6401	946.3959
18.58 + 2.34 x 17.17 is	359.1964 +	946.3959	1305.5923
2.34 − 13.46 x −19.84 is	220.6208 +	1305.5923	1526.2131
−13.46 + 0 x −32.68	439.8728 +	1526.2131	1966.0859
	1966.0859		

Note: The cumulative calculator method works only if the traverse is calculated in that direction and from that starting point which prevents the total row from registering a minus quantity.

Example: Using Method 1 above, the example shown in 16-2 of Davis' book *Elementary Plane Surveying* works out as follows:

Corner	*X*	*Y*			
A	300	300	A is 300 x (1200-400)	is	240,000
B	800	400	B is 800 x (300-600)	is −	240,000
C	1200	600	C is 1200 x (400-1000)	is −	720,000
D	1000	1000	D is 1000 x (600-1200)	is −	600,000
E	400	1200	E is 400 x (1000-300)	is	280,000
			Sum of minuses	is	1560,000
			Sum of pluses	is	520,000
			Double area	is	1040,000
			Area then	is	520,000

If coordinates are in feet, 520,000 square feet equals 11.94 acres.

Note: Because of the possibility of error in these calculations, the area should always be checked by triangles or planimeter. The method selected should be within the capacity of the calculating machine and be correlated with field methods.

Area Calculations for Segments of Curves

When the area of a parcel is affected by the existence of a road, the area of the roadway may be calculated and listed in the description and plat. For example: "containing a gross area of 49.77 acres, but after deducting a road area of 0.51 acre, leaving a net area of 49.26 acres." The area of private ways is seldom listed.

If lines of a parcel along a road are straight, the parcel area is determined by planimeter, DMD(s) or other methods. If a curve forms one or more lines of a traverse, and the area is required, it is determined by planimeter as follows: Plot the curve with its two radii on a large scale; superimpose the traverse thereon at the same scale, and planimeter the curved segments formed by the curves and survey courses.

If the area (for example the shaded segment ABC) must be found, more accurately, use one of the methods shown in textbooks. Perhaps the simplest one is to find the area of sector (AOBC) in the figure at left, then find the area of triangle (AOB) and subtract the two areas. (AO) and (BO) equal the radius (R), and angle (AOB) is the central angle (Δ). The area of the sector is found by proportion thus: sector area is to circle area as either (1) the ratio of angle (AOB) is to 360°, or (2) the ratio of length of arc (ACB) is to circumference of the circle. The area of triangle (AOB) is found by trigonometric formula. These calculations can sometimes be avoided by traversing the road or by tying in the road by triangulation. Traversing is better for streams and crooked roads because a small amount of field work can save a longer office calculation.

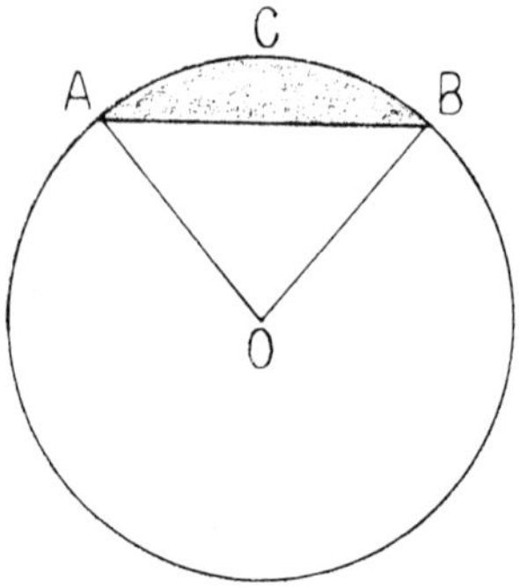

In choosing the method of area determination, remember that the acreage is usually the least significant factor of a survey. Descriptions prepared by the W. Va. State Road Commission containing curved areas often give the area as "about ——— square feet." If there is a large amount of curve work, it may pay the surveyor to obtain a copy of a book containing areas of curve segments and other curve data.

Balancing Calls with Cardinal Bearings

This problem was brought to the author's attention by Leon Mallow, a student in surveying at Glenville State College. When there are no cardinal bearings (N,E,S,W) in a traverse, the correction factor to balance Lats. and Deps. can easily be placed in the correct column. If the sum of the north Lats. is too high, the correction is subtracted; if the south Lats. are too low, the correction is added, so that the sum of the columns will be equal.

However, it sometimes happens that some of the calls are due north, east, south or west. In this event zeros are involved, and it may not be immediately apparent to which column the correction is applied. Assume a traverse with these calls, Lat. Deps. and corrections:

				Latitudes		Departures		Corrections	
Bearing	*Length*	*Cos*	*Sin*	*N+*	*S−*	*E+*	*W−*	*Lats*	*Deps*
North	500	1.000	0	500.0		0	0	0.84	3.16
East	400	0	1.000	0	0	400.0		0.67	2.53
South	600	1.000	0		600.0	0	0	1.02	3.79
N 76 W	400	0.242	0.970	96.8			388.0	0.67	2.53
	1900			596.8	600.0	400.0	388.0	By	
					-596.8	-388.0		Compass	
					3.2	12.0		Rule	

Davis & Foote page 455 state that the departure correction to a north line is to be added to the east departure because they were too small. Breed & Hosmer (1927, page 409 *Elementary Surveying*) say to apply to diminish the column that is too large and increase the column that is too small.

By inspection it is clear that 0.84 must be added to 500 to increase the north Lats. Similarly 1.02 is subtracted from 600 and the second 0.67 is added to 96.8. The problem is in which column to use the first 0.67. Adding the two corrected pluses gives 598.31 which is 0.67 shy of the 598.98 total of the corrected minus south Lats. Obviously the 0.67 should be placed in the north column so each column will add to 598.98.

For departures, it is obvious that the first 2.53 is to be applied to decrease the east Dep. and that the second 2.53 is to increase the west Dep. If the 3.16 were placed in the east column it would have to carry a minus sign in order to decrease the east total (which is the objective). So instead 3.16 is placed in the west Dep. column where by straight addition, it increases the west total. By the same reasoning 3.79 is placed in the west column. The sums of the two Dep. columns are now equal except for a rounding off figure of 0.01.

A rule seems to emerge here that no correction should appear in any column with a mathematical sign opposite from that at the head of the column; i.e. a correction of -3.16 should not appear in a column that carries a plus connotation. This appears to support the general rule.

The balanced Lats. and Deps. then for the above example are:

Latitudes		Departures		
N+	*S−*	*E+*	*W−*	
500.84			3.16	
0.67		397.47		
	598.98*		3.78*	*0.01 subtracted for final balancing
97.47			390.53	
598.98	598.98	397.47	397.47	

If 3.16 and 3.78 were placed in the east Dep. column with minus signs, the column totals would each be 390.53. In this case, the DMD and DA results are the same as when the totals are 397.47, namely the area = 216679.55 units.

Chapter 26

MONUMENTATION AND MARKING

Importance of Monumentation

The proper monumentation of corners and marking of lines is the highest priority job of the land surveyor. Courts have long given top priority to natural and artificial monuments. Markers established by surveyors must be so *unique* as not to be confused with any other object in the vicinity. The practice of using a stone without unique markings and witnesses is inexcusable.

Marking Corners. Although monuments are the "evidence" of corners, the two terms are generally used interchangeably in this book. Corners can be marked at the same time as offsetting or as a separate job. If there is a possible dispute between adjoiners, the marking should be delayed until agreement is reached. Stakes set on the true lines show all parties where the surveyor believes the line should be. It is best to eliminate disputes before the offsets are made. If adjoiners agree on corner location, they will not usually dispute the surveyor's location of a straight line between corners, even though a fence that may have been thought to be "on line" is found by the surveyor to be off here and there.

State laws or court decisions seldom establish standards for rural corner monumentation. Federal methods for public domain and National Forests provide an excellent guide for use. The following specifications are adapted from Chapt. IV 1947 *Manual,* USFS and traditional practices.

If an undisturbed monument is called for and found, it must be used. Additional markings may be placed upon a tree provided former markings are left undisturbed. If the surveyor must alter the condition of a monument, the evidence which identified its original location must not be destroyed. Complete notes must be kept of the description of the old monument and all changes and additions thereto. Reference the beginning corner of a survey by approximate bearing and distance from some well-known point such as the intersection of roads or streams.

Found Trees. Three hacks should be made on the tree on the side the line enters and leaves. These marks are usually placed at breast height but if there is danger of the tree being cut and removed, hacks may be made below stump height. Record in the notes the species of tree, its dbh, and what marks were placed on it. Old marks found thereon should have previously been recorded when it was first tied in.

Found Stones. A 2″-3″ letter should be chiselled in the top or side wherever it is flat. The grooves should be at least 1/8″ deep so they can be seen and felt and will not be confused with natural groovings. Unless the center of an "X" is needed to mark a more exact point, use some other letter, perhaps the first letter of the name of the client. Additional marks such as 1974 (year of survey) can be placed on selected stones if desired. Loose stones should be refirmed and surrounded with a mound of smaller stones. The dimensions and shape of the corner stone, the marks thereon, its depth in the ground, etc., should be recorded in the notes.

Other Types of Found Corners. Iron pipes, rods, gun barrels, axles, car springs, etc., have been used. Witness and tighten them. Ready-mix cement poured around them to below frost line is excellent. The notes must show what was found, size, and condition. Concrete monuments should be refirmed and description noted. Fence posts need not be marked because they lack permanence and stability. Points in unmarkable locations will be covered later.

New Corners. These should be marked and their descriptions recorded as described under "found" corners. If the spot is occupied by a tree of reasonable size and good species, it should be used as the corner. Saplings up to 3″ or 4″ in diameter whose crowns are not dominant are not good; a pipe or stone set beside them, or in their place, is better.

If the corner spot is in open ground, the surveyor has a choice: a stone marked by letter may be set in a hole, surrounded by a mound of stones. Unusual or symmetrical stones 20″ to 24″ long or larger are often found in the area. Use a flat-topped stone at least 4″ x 4″ on top. The hole should be deep enough—½ to ¾ the stone length if possible, to keep the stone in place and to show that a stone has been there (if later moved). Place the large end of the stone down for stability. It must be distinguishable from natural rocks nearby. There is nothing wrong with placing a bottle, or breaking a bottle, in a hole below a stone, anything durable and distinctive is good if it is described in the notes.

A pipe, or rod may be driven in the ground, perhaps set in cement, and surrounded by a mound of stones. New pipes should be 34″ or more long and 1″ or more in inside diameter, of galvanized iron. Other metal objects are good provided they are long enough and can be driven solidly below frost line. Pipes should project 6″-8″ above ground unless there is reason for less projection. Spikes may be used if ground does not allow deep driving. *Never* leave a wood stake for a corner.

If a fence post must be the corner, set a pipe snugly alongside. Reference the post well. Fence posts do *not* make good corners.

If the corner point falls on solid rock, an X or other letter can be chiselled in it. A mound of stones can be placed over the letter. If the corner is at some inaccessible point or a place where a corner would not long survive, a witness marker is needed. This should be established in one of the true lines running to the inaccessible point, as close to the point as practicable, but where it is most permanent. A witness corner could be a tree or monument on the bank of a watercourse, road, or railroad. If a road or right of way is involved, the witness should be on the edge of the r/w line. The witness is marked and described like a regular corner and should be referenced.

Monuments of any approved type may be set at intervals in boundary lines. They are good where a long line crosses a ridge or field. They should be fully described in the notes and in the survey description; they are legal markers just like corners. Monuments in fields or improved properties should be appropriate for the type of land use at that point.

Reference or Witness Markers

Nearly all corners can be destroyed by vandals, by passage of time, or by other acts of man. So it is essential to reference their location. The surveyor can perform *no more important service* than to evidence the location of corners as permanently as practicable.

Witness Trees. Trees make good references. They should be carefully selected according to these rules:

1. Be medium size, 6″-12″ dbh, and in the dominant canopy.
2. Be long-lived. Try not to pick two of the same species.
3. Be reasonably sound and capable of surviving competition.
4. Be on the client's property if possible.
5. Two trees to each corner, at right angles to each other if from the corner. Three trees for important corners.
6. Reasonably straight so measurements are clear.
7. Be the closest trees that are otherwise satisfactory.

Witness trees must be marked with 3 hacks, about 6″-8″ apart, at about breast height, facing the corner. The Hor. Dist. and bearing to the trees should be measured from the corner and recorded in the notes where the corner is described. The distance is always taken to the middle of the tree, because that point is least subject to change location. The chainman should stand beside and facing the bearing tree, looking at or right angle to the line of measurement, so as to more easily estimate the middle. If the tree leans toward or away from the corner, measure to a point 12″-16″ from the ground. Note vertical angle if slope distance is used.

Pointer Trees. In addition to witness trees, *pointers* should be marked with 3 hacks, facing the corner, if available within 10 feet or so of the corner. They are usually small trees or saplings. No measurement is made to pointers; the notes should say "marked pointers" or "dogwood and hickory pointers".

Other Witness Markers. A prominent cliff or boulder makes a good reference; a letter should be chiselled on the point selected, tied in by measurement. Other convenient objects should be referenced such as:

Specific corner of a building foundation
Point on culvert or bridge
A reference point or marker made by other surveyors
Fence corners
Center of a well
Electric or telephone pole, or other utility monument

Marking Lines

When. When offsets are made from random to true lines.

What Trees. The trees selected should usually be within 5 or 6 ft. of the line.

Type of Marks. Two hacks about breast height, 6″-8″ apart, made about horizontally, are placed on center-line trees, on the side the line of sight enters and the side it leaves. Hack through the bark into the wood. The same kind of marks are placed on trees within 5 or 6 feet of the line, on the side of the tree *facing the line.* Such marks will last the life of the tree. Notches are less desirable because it takes more time to make them and disease can enter the tree. The surveyor may select some other combination of hacks or blazes. Marks on objects other than trees will have to be considered carefully; do not place marks where they would be objectionable to the owner or to the public.

Other Boundary Markers. Some owners erect paper, wood or metal signs to denote the boundary, especially at key spots like road crossings. Tacks or nails should not go through the bark; if they do, the tree will grow around the nail (and metal sign).

Procedure. One man with a handaxe stands directly over the first station; a second man stands over the second station. They should see enough of each other to determine the direction of a straight line between them. If the line of sight is obstructed by a tree, it is a "center-line" tree. Each man notes such trees and discusses them with the other. Each man also preserves in his mind a "picture" as to which side of certain doubtful trees the line runs. Whoever is in charge signals for marking to begin.

The man at station 1 moves toward station 2, walking on the line of sight. As each markable tree is reached, he marks it, trying to remain standing on the line so as not to lose sight of the second man or "picture".

Obstacles. Brush, trees, boulders or the curve of a hill may prevent the men from seeing each other. In this case, one man or both may hang a cap or kerchief on an arm, rod, or straight stick, and raise it directly over the station, sometimes waving it slowly. If this does not work, try a method described under random line running, or perhaps a variation of the "picket" method.

Painting

A man with brush, paint and scraper should follow the man who marks the line. Three men (offsetter, marker, and painter) can, in good going, cover 1-1¼ miles per day. A horizontal paint band about 3″ x 6″ should be placed on or just above most or all the hacks, whether they are on sideline, centerline, corner or witness trees or pointers. Orange or yellow implement enamel is good. The paint should not be so thin as to run. Too much is better than too little. The tops of stones, pipes, etc., should be painted. Do not paint buildings and similar objects.

If painting is a separate project, one man can locate the hacks and scrape roughbarked trees (and help paint), while the second man paints. Extra paint can be carried in a knapsack in small plastic or polyethylene containers. Do not carry over ¾ gallon of paint in a gallon can due to danger of spilling.

One gallon should mark 1 to 1½ miles of new line; it may go farther on a repaint job. On one job of 8.65 miles, using 3 men one day and 2 men on three days, the average cost was $21.85 per mile, itemized as follows:

labor	– $18.56
mileage	– 1.90
paint	– 1.39
	$21.85

One gallon covered 1.8 miles of line. The 3-man crew cost $17 *per mile* and the 2-man $25. There was little non-effective foot travel on this project. Don't use cheap paint.

What Is a "Steel Pin" or an "Iron Stake"?

What do you expect to find when it is called for in a deed? A pipe? A rod? An axle? A gun barrel? A bolt? Isn't it time for the stake and pin mystery to be cleared up? How? By "telling it like it is". Describe uniquely the marker you set or find so it can't be mistaken.

Why not say: "set a 1″ x 36″ black iron pipe flush" or "set a 4 x 6 x 24″ stone marked ——— in a mound of stones" or "found a rusty ½″ iron pipe, replaced same by a 6 x 6 x 24″ concrete marker flush" or "set a 5/8″ x 34″ reinforcing rod with aluminum cap flush, marked John Doe LLS 683 1974"?

A landowner is entitled to an unbroken chain of title. The surveyor should provide an unbroken chain of unique marker descriptions.

Chapter 27

PLATS

Value of Plats

A plat of survey has long been considered the chief record "trademark" or handiwork of the land surveyor. A plat is high on the list of legal evidence used to locate property. It is usually the direct and unchanging work of the surveyor, whereas descriptions have ways of being changed by accident or intent as the years pass.

Occasionally a client, wishing to save money, may not want a plat. The surveyor however, should insist on furnishing a plat, which can save the owner and future owners much expense later. The laws of some states, including West Virginia, require that plats be made of subdivisions, and recorded before any lots are sold or offered for sale; plats of other land surveys must be recorded by the client or owner before or at the time the title to the property is conveyed. Both the West Virginia Code and Federal Regulations provide that coal mines shall be mapped. But even though not recorded, an identifiable plat if called for in the record, is legal evidence.

It is possible to make a complete and legal survey without making a plat. Everything that could go on a plat could be placed in a prose description. However, this would mean a long, unwieldy and confusing description, and there would be a tendency to forget or discard significant items. Even the resurvey of a city lot should include a new plat because of dimensions and markers differing from the original subdivision plat.

The Work Plat

A preliminary plat should be made, from which the final plat may be copied. It should contain all the data that goes on the final plat although abbreviations are usable. The best arrangement for the final plat can often be learned from poor arrangement of the preliminary plat. Plats should be the smallest size that can clearly show the data. Surveyors frequently make plats much too large. An 8½″ x 11″ size is good because:

1. It is recordable by photocopying processes with no size reduction and with a low recording fee.
2. Copies may be made on most photocopiers.
3. They are convenient for folding, mailing, attaching to deeds and filing. They need not be rolled or shipped in a tube.
4. Lettering can be done on most typewriters.
5. Larger plats can be made in two or more sections.

Scales of Plat.

If larger plats are required, give careful consideration to their recordability in the particular county involved. Scales of 10, 20, 40, and 50 or multiples thereof are preferred. Laymen seldom have any way of measuring a 30 or 60 scale plat. Consider a scale similar to the Assessor's plats; for example in West Virginia, 400 feet per inch for rural plats and 100 feet per inch for urban plats. If the Assessor's scale is not convenient, consider a small insert at the Assessor's map scale. This eliminates the need for tax mappers to replot calls at another scale.

Contents of Plat.

The West Virginia State Board of Examiners has promulgated rules and regulations covering what material should be placed on plats. Section 9 of those rules is shown here:

Section 9—Plats. Plats shall contain the following data:

9.01. State, County, and District where the land is located. If applicable, the City, Subdivision, Addition, Block and Lot number.

9.02. Deed Book, plat book, page number and other place where the deed, plat or other instrument covering the land, is recorded or found.

9.03. Names of adjoining owners with lot numbers, deed book and page numbers or other place where their deed, plat, or other instrument is recorded or found.

9.04. Overlaps, encroachments or gaps.

9.05. Type of corner and reference markers. For city lots, ties to edge or center of three nearest streets. For subdivisions, ties to two markers of each adjoining subdivision and to public triangulation stations in the vicinity, if present.

9.06. For rural and subdivision surveys, the magnetic and true north arrows and the amount and direction of the declination as of the time of the survey. If declination for rural surveys is determined at a place other than at the land surveyed, state place and date. For city lot surveys, show north arrow as on previous plat.

9.07. Bearings, distances, curve data (central angle, radius, arc and chord) and other measurements shall be shown with an accuracy consistent with the field equipment, methods and precision employed. Show comparative conflicting data if significant, in parentheses or other way. Enough dimensions shall be shown to enable all omitted dimensions to be determined by inspection or calculation.

9.08. Bearings shall be clockwise if practicable. Outside line bearings of subdivisions should be from the true meridian.

9.09. Scale of the plat showing inch first and its equivalent last.

9.10. Area of the entire tract and of any significant parts. Area of city lots if shown, in square feet. Total subdivision area shall include streets, alleys, and other non-lotted space. Accuracy of area figures shall be consistent with field methods employed.

9.11. Significant cultural, topographical or other relevant features. These may include buildings, utility facilities, pipelines, easements, fences, walls, driveways, watercourses, certificates with prescribed wording, etc.

9.12. Sheet and parcel number of tax map where land is located.

9.13. Location of land in relation to well-known landmark.

9.14. Date of survey, being as specific as practicable.

9.15. Date of plat if relevant; name, address, signature, and seal of the surveyor in charge. Certificates that may be required.

Here are some additional comments on plats:

Title.

The caption or title is usually placed in the top center or lower right corner, but may be placed at any other more suitable place. In Public Land states, the meridian, township, range, section and smaller parts may be included in the caption.

Some surveyors show on a plat "to be conveyed to (*name of proposed purchaser*)". This is considered unsound because the transfer may not be consummated and the surveyor should not anticipate an action that may not take place. The surveyor's duty is to make a survey of a selected tract of land and to describe the survey. Whether it is conveyed or not depends on matters beyond his control. The caption of course may contain either the name of the landowner or a (prospective) purchaser particularly if the purchaser is paying for the survey. It is necessary to have some name on the plat so it may be properly indexed in the public records. The report is the proper place to mention any proposed sale.

North Arrows.

Although there is no inflexible rule, the arrow of the orientation base should usually be parallel to the sides of the plat and be pointing up. The other arrow should be platted reasonably close to the declination angle and on the proper side. Fancy arrows only cost the client more money. The author prefers a half arrow head for magnetic north, the head being on the side away from the declination, and a double arrow head for true north. The specific point where declination is read should be noted on all plats, so a surveyor many years in the future can return to that point and determine the variation.

Scale.

A bar scale is often shown on plats in addition to the conventional

"1" = ———". This is a luxury rather than a necessity. The chief argument for a bar scale is to be better able to scale distances if distortion has occurred. However, distortion can occur at right angle to the bar scale, and can easily be determined by measuring several lines on the plat whose length is given.

Direction of Calls.

Most authorities agree that calls or lettering on a plat should be readable when the bottom or right side of the plat is next to the body. Some attorneys prefer calls to be readable as the plat is rotated 360 degrees in front of them; this is bad practice.

Monuments.

Traditional abbreviations such as IP for iron pipe may be used. Tree names are often shortened such as WO for white oak; these are subject to regional traditions.

Witness Markers.

It is not always feasible to show all witnesses on the plat. Witnesses for the beginning corner should be given first priority; they can often be placed in a circle enlargement in a blank part of the plat.

Adjoiners.

Show the line *between* adjoiners themselves with 2 or 3 dashes in the approximate proper direction.

General Location.

This may be shown by an insert sketch, by an arrow and miles to a landmark, or by latitude and longitude.

Border.

A solid border line or two will improve the appearance and readability of a plat.

The Final Plat

Final plat data is traced or taken from the "work" plat. The selection of paper, film, pencil hardness, type of pen and other drafting methods will not be discussed here because of their wide range and personal choices. Each surveyor will use the tools and materials that will best suit his needs and fit best with the copying processes to be used. However, the lines of the property should be the heaviest whereas lines of ties and between adjoiners, etc. should be lighter for contrast.

It should be emphasized that the original final plat should be retained in the file of the surveyor, and only copies furnished to clients and other parties. The copies must be good enough to be readable when placed in the public record.

Several typical plats are included here. A suggested form for rating plats is shown; it was developed by the author and John Campbell of Parsons, W. Va.

PLAT RATING FORM

Exhibit ____________

Item	*Rural*	*City Lot*	*Subdivision*	*Underground*
State, County, District, City	5	5	5	5
Subdivision, block, lot		5	5	
Deed book and page coverage	5	5	5	
Plat book		5		
Names of adjoiners	3	3	3	2
Deed book & page of adjoiners	2	2	2	5
Mineral ownership lines				5
Size and type of markers	10	10	10	5
Ties	4	4	4	
Coordinates				5
North arrow		5		5
Arrows, declination, date, place	5		3	
Consistency of measurements	3	3	3	
Bearings clockwise	2	2		
Bearings true			5	5
Scale	3	3	3	5
Area of total	5	3	5	
Area of parts	5	3	5	
Culture, topo, other items		5	3	8
Tax map and parcel number	3	5	3	
Landmark reference	5	2	2	
Year of survey	5	5	3	
Date revised				5
Names & signature of surveyor	4	4	4	5
Address of surveyor	1	1	2	
Seal or authority of surveyor	5	5	5	5
Elevations				5
Entries, pillars, etc.				5
Air currents				5
Abandoned portions				5
Outcrops				5
Clarity	10	10	10	5
Neatness	5	5	10	5
Total weight	100 ___	100 ___	100 ___	100 ___
Applicable weight & score				

Note: Estimate closeness of each item on plat to applicable weight and enter that figure after the weight.

Name of Judge ____________ Date ____________

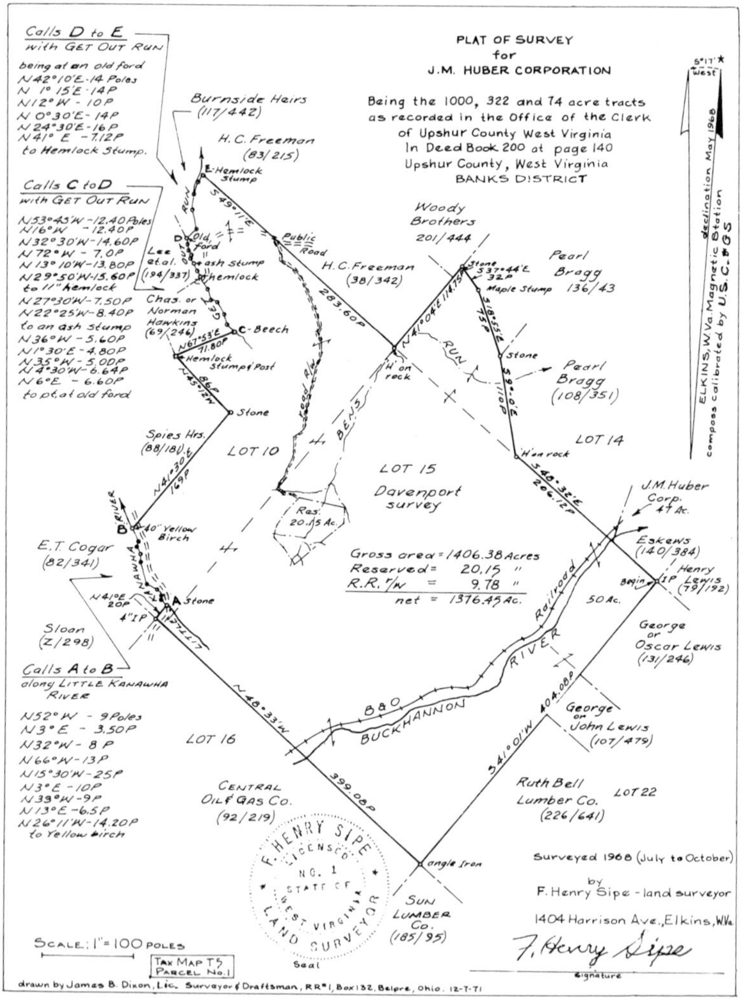
PLAT OF SURVEY
for
J.M. HUBER CORPORATION
Being the 1000, 322 and 74 acre tracts
as recorded in the Office of the Clerk
of Upshur County West Virginia
In Deed Book 200 at page 140
Upshur County, West Virginia
BANKS DISTRICT
Calls D to E
with GET OUT RUN
being at an old ford
N42°10'E-14 Poles
N 1° 15'E-14P
N12°W - 10P
N 0°30'E- 14P
N24°30'E-16P
N41° E -7.12P
to Hemlock Stump.
Calls C to D
with GET OUT RUN
N53°45'W-12.40 Poles
N16°W -12.40P
N32°30'W-14.60P
N72°W - 7.0P
N13°10'W-13.80P
N29°50'W-15.60P
to 11" hemlock
N27°30'W-7.50P
N22°25'W-8.40P
to an ash stump
N36°W - 5.60P
N1°30'E-4.80P
N35°W - 5.00P
N4°30'W-6.64P
N6°E - 6.60P
to pt. at old ford
Calls A to B
along LITTLE KANAWHA RIVER
N52°W - 9 Poles
N3°E - 3.50P
N32°W - 8 P
N66°W-13P
N15°30'W-25P
N3°E -10P
N39°W-9P
N13°E-6.5P
N26°11'W-14.20P
to Yellow birch
Burnside Heirs (117/442)
H.C. Freeman (83/215)
E-Hemlock Stump
S49°11'E
Public Road
283.60P
H.C. Freeman (38/342)
Woody Brothers 201/444
Stone
S37°44'E 32P
Maple Stump
Pearl Bragg 136/43
S18°55'E 75P
Stone
Pearl Bragg (108/351)
S9°0'E 110P
H on rock
LOT 14
S48°32'E 206.12P
N41°04'E 114.75P
H on rock
RUN
BENS
D Old ford
Lee et.al. (194/337)
ash stump
hemlock
Chas. or Norman Hawkins (69/246)
C-Beech
N67°53'E 71.80P
Hemlock Stump & Post
N45°12'W 86P
Stone
Spies Hrs. (88/181)
N41°30'E 169P
LOT 10
LOT 15
Davenport survey
Res. 20.15 Ac.
Gross area = 1406.38 Acres
Reserved = 20.15 "
R.R. R/W = 9.78 "
net = 1376.45 Ac.
B RIVER
40" Yellow Birch
E.T. Cogar (82/341)
KANAWHA
N41°E 20P
A Stone
4"IP
LITTLE
Sloan (Z/298)
J.M. Huber Corp. 47 Ac.
Eskews (140/384)
Begin IP
Henry Lewis (79/192)
50 Ac.
George or Oscar Lewis (131/246)
Railroad
RIVER
B&O
BUCKHANNON
N48°33'W
399.08P
S41°01'W 404.08P
George or John Lewis (107/479)
LOT 16
Central Oil & Gas Co. (92/219)
Ruth Bell Lumber Co. (226/641)
LOT 22
angle Iron
Sun Lumber Co. (185/95)
F. HENRY SIPE
LICENSED
NO. 1
STATE OF WEST VIRGINIA
LAND SURVEYOR
Seal
Surveyed 1968 (July to October)
by
F. Henry Sipe - land surveyor
1404 Harrison Ave., Elkins, W.Va.
F. Henry Sipe
Signature
5°17' West
declination May 1968
ELKINS, W.Va. Magnetic Station
compass calibrated by U.S.C.&G.S.
SCALE: 1" = 100 POLES
TAX MAP T5
PARCEL No. 1
drawn by James B. Dixon, Lic. Surveyor & Draftsman, RR#1, Box 132, Belpre, Ohio. 12-7-71

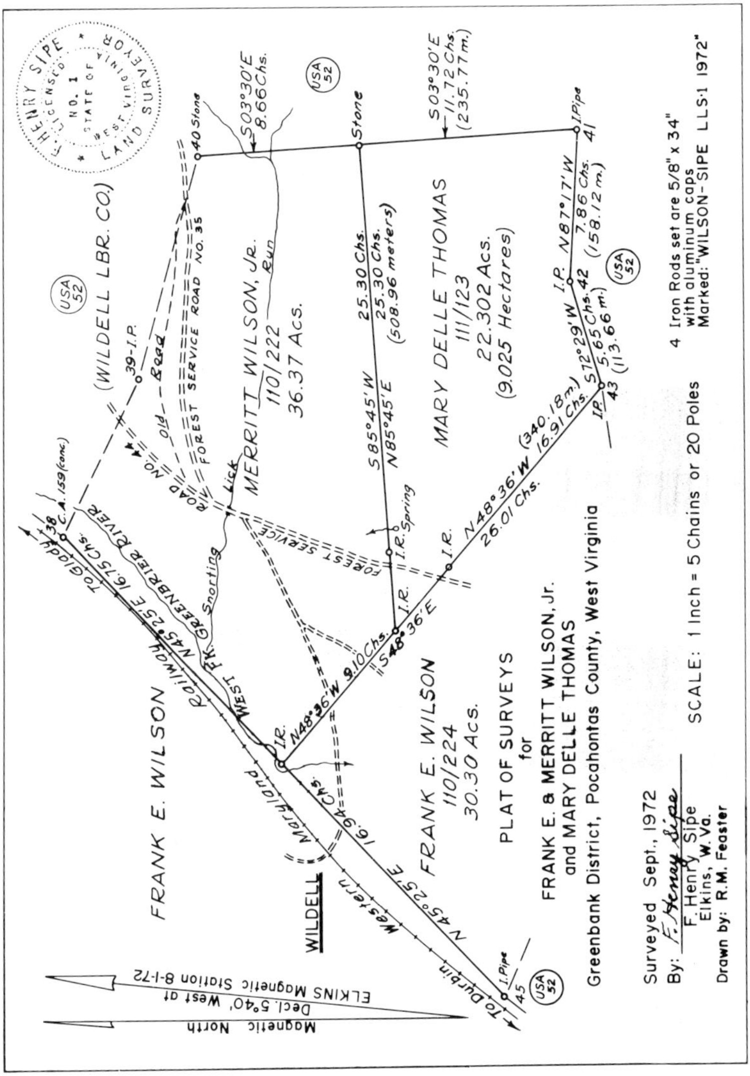
PLAT OF SURVEYS
for
FRANK E. & MERRITT WILSON, Jr.
and MARY DELLE THOMAS
Greenbank District, Pocahontas County, West Virginia
SCALE: 1 Inch = 5 Chains or 20 Poles
Surveyed Sept., 1972
By: F. Henry Sipe
F. Henry Sipe
Elkins, W. Va.
Drawn by: R.M. Feaster
4 Iron Rods set are 5/8" x 34"
with aluminum caps
Marked: "WILSON-SIPE LLS-1 1972"
Magnetic North
Decl. 5°40' West at
ELKINS Magnetic Station 8-1-72
F. HENRY SIPE
LICENSED
NO. 1
STATE OF WEST VIRGINIA
LAND SURVEYOR
FRANK E. WILSON
(WILDELL LBR. CO.)
MERRITT WILSON, JR.
110/222
36.37 Acs.
MARY DELLE THOMAS
111/123
22.302 Acs.
(9.025 Hectares)
FRANK E. WILSON
110/224
30.30 Acs.
WILDELL
Western Maryland Railway
To Durbin
To Gladys
N45°25'E 16.94 Chs.
N45°25'E 16.75 Chs.
GREENBRIER RIVER
WEST FK.
Snorting Lick Run
FOREST SERVICE ROAD NO. 35
FOREST SERVICE ROAD NO. 44
Old Road
C.A. 159 (conc.)
38
39-I.P.
40 Stone
41 I.Pipe
42 I.P.
43 I.P.
45 I.Pipe
Stone
I.R.
I.R. Spring
S03°30'E 8.66 Chs.
S03°30'E 11.72 Chs. (235.77 m.)
S85°45'W 25.30 Chs.
N85°45'E 25.30 Chs. (508.96 meters)
N48°36'W 9.10 Chs.
S48°36'E
N48°36'W 26.01 Chs.
N48°36'W 16.91 Chs. (340.18 m.)
S72°29'W 5.65 Chs. (113.66 m.)
N87°17'W 7.86 Chs. (158.12 m.)
USA 52

PLAT OF SURVEY
of
Land conveyed by Maggie C. Bent to Laura Bent White
by deed dated March 2, 1961 recorded in the Office
of the Clerk of Randolph County, West Virginia
in Deed Book 222 at page 591.

13900 square feet in the City of Elkins

SCALE: one inch equals 40 feet

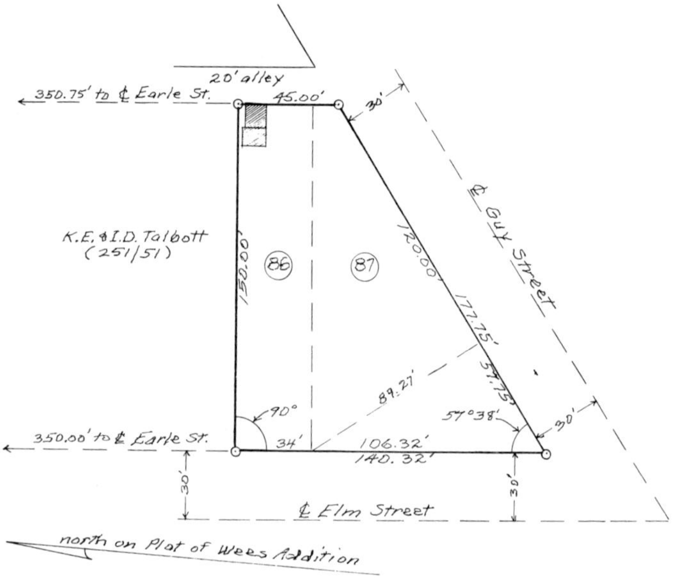

NOTE: at each corner is an iron rod
with aluminum cap marked

BENT 1969 SIPE

Surveyed August 1969
by
F. Henry Sipe
Land Surveyor
Elkins, W.Va.
LICENSE APPLIED FOR
May 26, 1969
F. Henry Sipe

Rev. 10/7/69

Chapter 28

DESCRIPTIONS

Probably no other office job is done so poorly by some surveyors as the writing of descriptions. The reason is that attorneys have long been the writers of descriptions, and surveyors had little training in that work. The surveyor would give only a plat to the client, and the attorney or even the landowner would try to interpret the plat and write the description.

The author believes that the surveyor is the best qualified person to interpret his own plat and field notes, and should be solely responsible for writing the description. However, any surveyor should welcome suggestions from an attorney when some legal aspect of the description is involved. Sometimes the changing of a word or two can change the intent of the instrument as to what is to be conveyed, or the ease with which the property can be resurveyed many years hence.

There are several good texts on writing descriptions and the author will not go into detail on this subject. However, the description must clearly identify the location of the property so that its boundaries can be established in the future. The beginning corner should be the one most certainly identified. All data on the plat should be included plus details not susceptible of being put on the plat. The surveyor should study the form of descriptions in a new area and try to conform to the traditions there.

The following items should be included (or purposefully rejected) for new or revised metes and bounds descriptions. They are in three groups to conform to general West Virginia tradition:

The first paragraph (the Preamble) includes the *general* description:

1. "A tract (parcel or lot) of land in".
2. District, County and State. (City if desired.)
3. Watershed and/or tributary if appropriate.
4. Approximate distance and direction to a well-known landmark.
5. "more particularly described as follows:"

The body of specific descriptive items then follows in tabular form, with each Bearing starting a new paragraph:

6. "Beginning at ---" (what marker, tree, pipe, etc.; size, species, how marked, whether found or set, etc.).
7. Its environment (at fence corner, on streambank, ridge, etc.).
8. Corner to whom, or in line of whom (list in order you run).
9. If an interior lot, or a tie to an outside corner(s) is required, show ties here.

10. Bearing and distance to nearby references (trees, buildings, etc.).
11. "thence with ———" (adjoiners, cultural features such as edge or meanders of a stream, road, or with or along a fence, etc.).
12. Bearing (this starts a *new* paragraph).
13. At specific distances, passing what markers, new adjoiners, objects like crossing centers of roads or streams.
14. "in all ———" (give *total* length of line).
15. To what marker or monument (see beginning items 6 through 11).
16. If a curve is involved, state radius & length. (For convenience of later resurveys, bearing and length of long chord can be listed.)
17. After listing all calls, state "containing ——— acres more or less" (include square footage only for small lots) "as surveyed ——— (month and year) by ——— (name and address), "and as shown on a plat attached hereto and made a part of this description."
18. If bearings are other than current year's magnetic, give basis here.
19. Marking and painting of lines should be mentioned here, if done.

Now comes the closing or "being" clause:

20. "Being the same land (or part of) conveyed by ——— (grantor(s)) to ——— (grantees(s)) by deed dated ———, recorded in the Office of the Clerk of ——— County West Virginia in Deed Book —— at page ——."
21. Additional references required, such as Plat Book, Will Book, etc.
22. Date description was written; name, address and Seal of surveyor.

Undesirable words in descriptions: stake, iron pin, reversed, near, about, down the hill, up the creek.

Several descriptions are shown in this chapter including an old one from New Hampshire furnished by the late Garald Hyde. A suggested form for rating metes and bounds descriptions is also included.

Partial Description of Survey for J. M. Huber Corporation

A tract of land in Banks District Upshur County, West Virginia, on waters of Right Fork of Buckhannon River and Little Kanawha River about two miles southwest of Selbyville, more particularly described as follows:

Beginning at a 1¼″ iron pipe found at fence junction, known as the "Comp Lewis" corner, corner to C. W. Eskew, Henry Lewis and

George Lewis, from which an 8″ lynn is N 86 W 23.4′ and a 7″ red maple is S 18 E 6.7′ thence (calls omitted)

S 48-32 E, at 135.7 poles crossing said Railroad, at 138.8 poles crossing said River, at 140.9 poles passing 6.3′ SW of 1½″ iron pipe, sugar maple called for but only root hole found, then with Eskew, in all 206.12 poles to beginning, containing 1406.38 acres, but after excluding 20.15 acres reserved for a pond, and 9.78 acres for a 60′ railroad r/w, leaves a net of 1376.45 acres more or less, as surveyed in 1968 by F. Henry Sipe of Elkins, W. Va., with opensight compass with 4″ needle with error of closure of 1 in 3480, as shown on a plat attached hereto and made part of this description; all lines and markers being marked and painted yellow;

Being the 1000, 322 and 74 acre tracts conveyed by T. E. Shaw to J. M. Huber Corporation by deed dated 12/26/67 recorded in Office of the Clerk of Upshur County West Virginia in Deed Book 200 at page 140.

Description of Survey for Mary Delle Thomas

A resurvey of a tract of land of 22.26 acres shown as (C) on a plat of E. F. Smith PE, dated May 31, 1963, conveyed by Merritt Wilson Jr. et al. to Curtis E. Wilson by deed dated June 21, 1963, and conveyed to Mary Delle Thomas by deed dated December 16, 1963 recorded in the Office of the Clerk of Pocahontas County, West Virginia, in Deed Book 111 at page 123, more particularly described as follows:

Beginning at Corner 43 of U. S. Tract 52 and Corner of Frank E. Wilson land, a 1″ iron pipe filled with cement, found on a ridge, in a mound of stones, painted red, from which a 12″ cherry is S 78¼ E 15.5 feet, a 15″ cucumber is N 25¼ W 5.3 feet (original witness) and an 8″ red maple is S 52½ E 9.0 feet, thence with Frank E. Wilson land,

N 48-36 W, at 12.56 chains passing a 5/8″ x 34″ iron rod with aluminum cap marked *Wilson-Sipe LLS 1 F. Logan 1972* from which a 4″ Norway Spruce is S 32-40 W 8.9 feet, a 3″ Norway Spruce is N 30-30 E 15.4 feet and a 4″ Norway Spruce is S 72-05 E 24.3 feet, at 13.11 chains crossing the public road to Durbin, in all 16.91 chains (340.18 meters) to a corner of Merritt Wilson Jr. set a 5/8″ x 34″ iron rod with aluminum cap marked *Wilson-Sipe LLS 1 1972,* from which a 20″ spruce is N 48-36 W 2.64 chains, thence with Merritt Wilson Jr. land.

N 85-45 E, at 3.55 chains crossing said public road, at 4.00 chains set a 5/8″ x 34″ iron rod with aluminum cap marked *Wilson-Sipe LLS 1 1972* from which a 5″ hawthorn is S 8-40 E 11.0 feet, a double 10″ apple tree is S 68-00 E 25.3 feet and a 2″ spruce is N

68-05 E 44.3 feet, at 5.33 chains passing about 25 feet north of a spring, in all 25.30 chains (508.96 meters) to a point in line of U. S. Tract 52, but 10.7 feet east of a red painted line, set a 7 x 11 x 22″ stone in mound of stones from which a 10″ black cherry is N 22 E 17.8 feet, a 13″ spruce is S 52-15 W 26.1 feet and a 20″ black birch with red paint is N 64-50 W 12.1 feet, marked said stone with a W, thence with said U. S. line but easterly of the red paint

S 3-30 E 11.72 chains (235.77 meters) to Corner 41 of U. S. Tract 52, found a 1″ iron pipe filled with cement, in a mound of stones on a ridge, from which a dead 13″ beech snag is S 14¼ E 14.5 feet and an 8″ beech is S 50 W 16.5 feet; made a new witness a 7″ beech S 26-45 E 12.2 feet, thence with the ridge and U. S. line

N 87-17 W 7.86 chains (158.12 meters) to corner 42 of U. S. Tract 52, found a 1″ iron pipe in a mound of stones; bearing trees not recorded, thence

S 72-29 W 5.65 chains (113.66 meters) to the beginning, containing 22.302 acres (9.025 hectares) more or less, as surveyed by F. Henry Sipe September 1972 and as shown on a plat attached hereto and made a part of this description;

All bearings are magnetic; the SW and N lines and all markers and witnesses have been marked and painted.

Description of Survey

A parcel of land in Block 10 of the Wees Addition to the City of Elkins, Randolph County, West Virginia, more particularly described as follows:

Beginning at the northeast intersection of Guy and Elm streets, 30 feet from the center of said streets as indicated by curb lines, set a 5/8″ x 34″ iron rod with aluminum cap, from which the edge of an old brick walk is 1.35′ southerly and 0.80′ westerly, thence with the edge of Elm Street

Northerly 140.32 feet to a point 30 feet from the center of Elm Street, 15.02 from the edge of the curb, and 0.30′ from edge of said walk, set a 5/8″ x 34″ iron rod with aluminum cap, thence with a line of that part of Lot 86 heretofore conveyed away, and at a right angle to Elm Street

Easterly 150.00 feet to a point on the west edge of an alley, set a 5/8″ x 34″ iron rod with aluminum cap, thence with the west edge of said alley,

Southerly 45.00 feet to a point 30 feet from the center of Guy Street, and 0.70′ from the edge of the end of said walk, set a 5/8″ x 34″ iron rod with aluminum cap, thence with the northwest edge of Guy Street

Southwesterly 177.75 feet to the beginning, containing 13,900 square feet more or less, as surveyed by F. Henry Sipe of Elkins and as shown on the plat attached hereto and made a part of this description; all aluminum caps were marked *Bent 1969 Sipe:*

Consisting of three parcels: (1) the residue of Lot 86 of Block 10 of the Wees Addition to the City of Elkins as shown on a Plat of said Addition in Deed Book T, page 52 in the Office of the Clerk of Randolph County, West Virginia; (2) Lot 7 of Block 10 of said Addition, and (3) a triangle of land awarded to Maggie C. Bent in a lawsuit 69 W. Va. 1911, page 11; all three parcels having been conveyed by Maggie C. Bent to Laura Bent White by deed dated March 2, 1961 recorded in said Clerk's Office in deed book 222 at page 591.

Volume 253 Page 554 Hillsborough County Records

KNOW ALL MEN BY THESE PRESENTS,

THAT I, Mary Leach of Goffstown in the County of Hillsborough and State of New Hampshire, Widow of Samuel Leach, in consideration of One hundred dollars to me paid by Margaret Leach of said Goffstown Widow of Joseph Leach the receipt whereof I do hereby acknowledge, have given, granted, bargained, sold and conveyed and do for myself and my heirs, by these presents, give, grant, bargain, sell and convey unto the said Margaret & her heirs and assigns, forever,

The following tract of land situated in said Goffstown and bounded as follows: commencing on the north line of a tract of land the said Samuel Leach deeded to said Joseph by deed dated November 25, A.D. 1844, Recorded Hillsborough Records Volume 230, Page 146, at a point south of the partition in the house which is east of the west front door; thence following said partition north at the east side of the stairs till it comes to the north room at the head of the Stairs; thence northerly on the west side of said north room to a point north of said partition and four feet distant from the house; thence east 4 feet distant from the house to the east end of the house; thence south by the east end of the house till it strikes the north end of the land conveyed in the deed first named; and thence west on siad [*sic*] line to the point of Beginning, reserving the right for myself and my heirs forever to pass through the room at the head of the Stairs to the front Parlor or square room over my Ketchen [*sic*] and also reserving the right so long as I the Grantor shall live to go through the premises conveyed to get water from the well and wood from the wood shed and to get to the hog pen. I also convey to said Margaret three Bed rooms which are situated in said house, west and north of the front room above described which I the Grantor have above reserved the right to get to by passing through a room above conveyed and reserving to my heirs & assigns the right to pass through the premises conveyed to get to said room. I also convey to said Margaret one undivided half of the Wood shed hog pen and back-house which stands at the east end of the house above conveyed together with one undivided half of the land upon which said shed hog-pen & Back-house stands.

Is It Still Legal?

An attorney checking real estate titles for the Right of Way Division ran across the description below, which was contained in a deed dated 1951.

"All of a certain peace or parcell of real estate, with general warnty of title, the following described real estate to wit.

beginning on a iren stake at the lore side of the state highway, thence with the said road to a stone, a corner of ———, thence with the said ——— line to a stone near the drean thence with said drean to the main branch to a painted rock at a wire fence, a line comily called the ——— line thence north to a irn stake, thence north to the irn stake at the beginning suppose to contain 1½ (one and onehalf) acors more or less."—W.VA. SURVEYOR

Land Surveyors Should Write Deed Descriptions

The following letter was written by William Youngblood, Professional Engineer & Surveyor, Lexow Ave., Upper Nyack, N.J. to the editor of Surveying & Mapping Division of the American Society of Civil Engineers:

"Gentlemen: I wholeheartedly concur with your thoughts to raise the standards of Land Surveying as high as possible, and it is to this end that I am now writing to you.

"As a self-employed Land Surveyor, I have been constantly bothered by the large number of faulty deeds drawn up by persons totally unfamiliar with the requirements of retracing a deed or preventing ambiguity in the description. As in most states only a licensed surveyor can prepare and file in the County Clerk's office a land subdivision map, I question why a deed which conveys or subdivides land is permitted to be prepared and filed by persons other than land surveyors.

"I proposed that we strive for legislation in each state requiring that certification of some type be prepared by a licensed land surveyor (either on the instrument to be recorded, or separately), indicating that the deed is correct and/or sufficient to allow the correct determination of the property."—Surveying & Mapping Division Newsletter (ASCE) June 1971

Metes and Bounds Description Rating Form

NOTE: If item does not apply, reduce maximum weight by that amount.	Weight	Rating
DESCRIPTION TYPEWRITTEN	4	
CAPTION: "Description of Survey of ———" (or for ———)	2	
PREAMBLE:		
District, County, State (City or Town)	3	
Watershed	2	
Landmark	2	
:more particularly described as follows:"	1	
BODY:		
Markers:		
Environment	2	
At corners: type, size, how marked	5	
At corners: found or set	5	
For witnesses: adequate in number and quality	5	
For witnesses: type, size, bearing, distance	5	
Along lines	3	
Adjoiners:		
At corners	3	
Along lines	2	
Lines:		
Bearings and distance clockwise	2	
Bearings and distance start new paragraph	4	
Bearings and distances consistent	1	
Curve data	2	
Proper meander calls	2	
Acreage: consistent with measurements; "more or less"	5	
"As surveyed by ———" (name and address)	4	
Date of survey (year or year and month)	2	
"As shown on plat attached hereto and made part of this description:	2	
Orientation from true meridian	3	
Lines marked and painted	5	
BEING CLAUSE: (previous instrument coverage)		
Grantor, grantee	2	
Date of deed	2	
Book and page in records	4	
Lot, block, subdivision	2	
Plat book reference	2	
Other title source or coverage	2	
ABSENCE OF UNDESIRABLE OR AMBIGUOUS WORDS	3	
Signature of Surveyor	2	
Date description written	2	
Seal imprint	3	
maximum	100	
applicable	——	

Chapter 29

OTHER RECORDS

The Report of Survey

The chief office products of the land surveyor, have been his plat and description. But one of his most valuable tools has generally been overlooked—a report (or certificate) of survey. Heretofore its chief use has been for lawsuits, dividing estates, or court-ordered surveys.

Basically, the purpose of a survey "report" is to record significant facts that are not shown on the plat or placed in the description. Plats and descriptions should not be unduly complicated. Listed here are items that should be included in a report of survey:

(1) Name of surveyor and date or dates of survey.
(2) Name and address of client or interested parties.
(3) What land was surveyed.
(4) What records were examined.
(5) Who were contacted and what information was obtained from them.
(6) Problems encountered and how they were resolved or treated.
(7) Encroachments, gaps, occupancy, possession, other unusual items.
(8) Signature, address, and seal of the surveyor.

Other items that can be placed in a report:

(1) Methods and equipment used.
(2) Error of closure or uncertainty of position.
(3) Distribution of records.
(4) Name and address of helpers.

Advantages of a Report of Survey:

(1) It records the weight given by the surveyor to the evidence studied.
(2) It can provide a list of deeds, etc. in various chains of title that are helpful in other surveys in the area.
(3) It makes these data available to the client and his successors, to the surveyor's successors, to the adjoiners, and, if recorded, to the public, all at a fraction of the cost of assembling it in the future.
(4) It gives assurance that an adjoiner's rights have been considered.
(5) It provides a handy way to refresh the memory of the surveyor.
(6) It reduces the chance of a surveyor being involved in negligence.
(7) It allows a resurvey to be made at less cost.
(8) It can help prevent or solve boundary disputes.

(9) It lends the finishing touch to what should be a *professional* job, and advertises the knowledge and skill of the surveyor.

Disadvantages of a Report of Survey:

(1) It opens the surveyor's knowledge and skill to the scrutiny of other surveyors and the public.

(2) It adds a few dollars to the cost of the current job.

A survey report can sometimes be used *before* the job is complete, to place before interested parties, the tentative conclusions and proposals of the surveyor. Replies from interested parties can be considered before the final report is issued.

When surveyors begin to use this valuable tool—the *Report*—they will find it of great help in being accepted as professionals by the public. Several reports or certificates of survey are shown in this chapter. (See page 174 for Report Rating Form.)

Report of Survey for J. M. Huber Corporation

I certify that I did, in 1968, resurvey for J. M. Huber Corporation, certain lands conveyed by T. E. Shaw to said corporation by deed dated 12/26/67, recorded in the Office of the Clerk of Upshur County, West Virginia, in Deed Book 200 at page 140; that the tracts surveyed were the 1000, 322 and 74-acre tracts and that these are contiguous:

That all lines and corner markers were marked and painted yellow in 1970;

That I wrote a description and made a plat of my survey, which are attached hereto and made a part of this report;

That I examined all the deeds referred to on the plat, and in most cases also examined prior deeds in the chains of title;

That I furnished nearly all adjoining owners with copy of a preliminary plat; that I evaluated all information obtained from them; and that I found no problems or encroachments except as noted herein;

That Mrs. Pearl Bragg complained verbally that the 74-acre lines were wrong, but wrote on 11/22/68 that she was satisfied with them; that the hemlock stump at "E" was in part located by a tie 128.73 poles to a 30″ yellow birch with old marks at base of a 12′ cliff, a Burnside corner; that the only encroachment is an area ½ pole by 5½ poles enclosed by fence at the north corner of the 74-acre tract;

That these tracts were resurveyed with 4″ needle open-sight compass, the error of closure being 1 in 3480; that I was assisted by Oscar Riffle of French Creek, Mark Riffle of Alexander, Frank Logan and James Logan of Elkins.

Report of Survey for Frank E. Wilson, Merritt Wilson, Jr. and Mary Delle Thomas

I certify that I did in the summer of 1972 make surveys for the above parties, of certain boundary lines of lands in Greenbank District, Pocahontas County, West Virginia, easterly of Wildell;

That I made plats and descriptions of said surveys, which are attached to and made a part of this report;

That I examined the deeds shown on said plat, and a plat attached to certain of said deeds bearing E. F. Smith's PE name; that I examined the records of the U. S. Forest Service Tract No. 52 (Wildell Lumber Company);

That no problems were encountered in these surveys except that certain distances and angles in the 1963 deeds proved to be slightly wrong; that the actual distance from U. S. Corner 39 to U. S. Corner 40 is 12.04 chains and not 11.04 chains; that the line between U. S. Corners 45 and 38 was given as 34.40 chains whereas it measured 33.69 chains, and that this deficit was prorated proportionately because of equal seniority of deeds;

That no encroachments by the property were noted; that the Forest Service red paint between U. S. Corner 40 and 41 encroaches upon the Merritt Wilson Jr. and Mary Delle Thomas parcels; that these matters are being referred to the Forest Supervisor in Elkins;

That the line between U. S. Corners 45 and 38 was surveyed by Wild 1′ transit, and the other lines with Warren-Knight open-sight Compass #17841; that the error of closure of the Mary Delle Thomas parcel was 1 in 1794; that I was assisted in these surveys by Frank Logan and James Logan of Elkins;

That two copies of the plat and description and one copy of the report are being furnished to Frank E. Wilson, Merritt Wilson, Jr., and Frederick L. Thomas, husband of Mary Delle Thomas, and that one copy of the plat, descriptions and report are being sent to the Forest Supervisor, U. S. Forest Service, Elkins, W. Va.

Certificate of Survey

I certify that I did, in August 1969, survey the land conveyed by Maggie C. Bent to Laura Bent White by deed dated March 2, 1961, recorded in the Office of the Clerk of Randolph County, West Virginia, in deed book 222 at page 591;

That before making the survey I examined these deeds:

April 2, 1894 from P. H. Wees to Maggie C. Bent for Lot 86 of Block 10 of the Wees Addition to the City of Elkins (50/432).

August 25, 1914 from Maggie C. Bent to Sadie Rogers for the northerly 20 feet of Lot 86 (106/95).

November 12, 1897 from P. H. Wees to Maggie C. Bent for Lot 87 in Block 10 of the said Addition (56/104).

That I examined the Plat of said Addition recorded in Deed Book T at page 52, and the lawsuit between Mrs. Bent and the City of Elkins, which she won on appeal to the Supreme Court 69 W. Va. 1911, page 11. In this suit, Mrs. Bent was awarded title to a triangle of land southwesterly of Lot 87 as shown on the Plat of Addition;

That I measured to all apparent lot lines in Block 10 and in the blocks between Earle and Prospect streets and between Guy and Cherry streets along Fayette; that I found only two markers in Block 10, one being a pipe said to have been set by Jake Goddin between Lots 80 and 81 on Elm Street, and the other being an old pipe apparently set by County Surveyor McLaughlin in 1959, at which time he "located and established" the north line of the Bent property. This pipe was 0.15′ south and 0.15′ east of my location of the NW Bent corner. The pipe between Lots 80 and 81 appeared to be 1.87′ too far south when measured from the center of Earle Street;

That there are a number of problems that make it difficult to adjust the discrepancies found in Block 10, namely:

1. Most significant is the poor original plat of the Wees Addition. It does not show the width of streets adjoining Block 10. These scale about 60 feet, which figure has been used on all subsequent maps.

2. The owners of Lots in Block 10 have not used the measurements shown on the Plat or have not started from the same base line. They have established lines and constructed buildings southerly of their proper locations if measured from the midpoint of Earle or Guy streets.

3. The Plat angle at Guy and Elm is less than I found it to be. This apparently helps create a 3.3 foot shortage in Block 10 along Elm Street. The shortage at the alley is about 1¼ feet but this cannot be checked against the plat because the original Plat shows no dimension for Lot 87 at the alley. If the 3.3 feet were deducted proportionately from all lots in Block 10, the Bent property would absorb about 1 foot of it.

There is a similar lack of consistency in the block between Earle and Prospect, but no apparent shortage of distance. There is a substantial surplus between Guy and Cherry along Fayette.

All this indicates that Guy Street is farther northwest than the original Plat indicates. Guy Street thus would appear to encroach on the Bent land. The original T/52 Plat indicates that the Bent land should cover nearly half of the old brick walk along Guy Street, but not cover the entire surplus in the adjoining block along Fayette Street.

As stated before, the lines of possession that have developed in the 79 years since the Addition was laid out, have crept southerly from their platted locations.

Two choices appear to be available to locate the north line of the Bent property:

A. One is to give priority to the possession lines of walks and several structures (particularly on the alley end of the lots) existing in the lots to the north of Bents. To do this would place an old shed on the back of the Bent lot about two feet on the adjoining lot. It would also ignore a cultivation line that appears to have existed for many years between Bent and the 20 foot strip sold off Lot 86. It would give to the other lots in Block 10 about 1.9 feet more frontage than their deeds call for along Elm Street (or that the original Plat calls for).

B. The other is to give priority to the cultivation line and line of shrubbery between Bent and Talbott and to take from Bent about 3.3 frontage on the Guy Street end, provided the present Guy Street location can be supported. It takes from Bent about 1¼ feet at the alley. This choice recognizes the right of the public in the old brick walk along Guy Street, which walk may have been built at the time of the lawsuit with the City.

The additional points in favor of choice B are:

1. There is exactly enough front footage in the other lots in Block 10 to equal their original Plat footages.

2. There is a slight excess of frontage on the alley.

3. Mrs. Bent was the first purchaser of a lot in Block 10 (Lot 86). 20 feet was sold off this lot in 1914 and the possession line could be at least 55 years old. I doubt that any of the possession lines farther up the block are that old.

I have therefore selected choice B for my marker locations and I believe they represent the superior evidence.

A surveyor's location or relocation of corners is always subject to change based on new and superior evidence. Until such evidence is discovered, I recommend the acceptance of corners as shown on my plat.

No other lot owners in Block 10 have been informed of my conclusions, which are shown on a plat and description of survey attached to and made a part of this certificate.

Report to Client

When the job is completed, a letter may be written to the client enclosing two (or more if desired) copies of the plat, description and report of survey. A list of costs, itemized if desired, is given. Any borrowed deeds or other data are returned. When there is a question about receiving payment for the work, some surveyors have effectively used the practice of withholding the records until payment is made or assured. This practice was authorized by law in Virginia well over 100 years ago. Sometimes a meeting is arranged when the records are delivered and payment made.

It is helpful to staple all records together in the upper left corner. When plats are larger than 8½″ x 11″, they should be folded from the right to left accordionlike, then from bottom to top. The upper left corner is then turned down so it is left outside the staple. In this manner all papers are conveniently positioned for examination and refolding.

Report to Other Interested Parties

It is usually good practice to furnish copy of the final plat to adjoiners and specially interested parties. Descriptions are optional. Discretion should be used in distribution of reports.

The Surveyor's File

The surveyor will ordinarily retain original plats and possibly the original description and report, furnishing only suitable copies to clients. A 9 x 12 manila folder may be used for each client or job, with identifying name or number on the tab. If there is more than one job per client, the location could be added. Place all data relating to the survey in the folder. When payment has been made, move the folder to a closed file. It is helpful to maintain an index map showing what tracts have been surveyed. The surveyor can check the map to refresh his memory for past surveys made in a new client's area.

Many a surveyor's files have been burned up after his death. When a surveyor becomes disabled, ceases to survey or passes away, it is desirable to see that his files are not destroyed. They might be disposed of to:

1. A company client
2. Another surveyor in the area, perhaps one who is taking over the former surveyor's business
3. A public depository
4. The current owner of the property, but this is a last choice.

Chapter 30

SURVEYING RURAL WOODLANDS WITH THE TRANSIT

By L. Bruce Sumner
Cadastral Surveyor, U. S. Forest Service
California Region

General

This book on compass surveys of rural land would not be complete without some discussion of transit surveys in the woods. There will be times when the transit should be used rather than the open-sight compass.

This chapter is not meant to be a comprehensive treatise on transit surveys; there are many textbooks that explain the use and care of the transit. Neither is it meant to explain all possible methods that can be used by rural surveyors, such as running by the transit's compass alone. What will be described is one very practical method that utilizes the inherent capabilities of the transit, and which, with reasonable care and proper chaining practices, will produce errors of closure of one part in 2000 to 5000. Regardless of what method the surveyor chooses, the following points should be kept in mind:

1. *Double all angles* to compensate for instrumental error, to obtain more precise readings, and to provide a running check.
2. *Avoid short backsights* of less than 50 to 100 feet. Short backsights can introduce errors that are carried throughout the survey, increasing in direct proportion to the distance covered.
3. *A survey that does not result in a closed geometric figure can not be absolutely relied upon.*

The Instrument

Before continuing further, it would be well to describe the type of transit best suited for woods work. The transit styles available are many and varied and each has its advantages for a particular type of work. For woodland work the transit should have these features: vertical circle or arc, erecting image telescope of low power, a telescope level, good compass with declination offset, extension leg tripod, and easily-read horizontal circle numbered both right and left.

The large engineer's transit will do the job, but is heavy, often graduated to less than one minute (which complicates reading) and usually has a telescope with too much power for woods work. The optical-reading transits are easily read, but unfortunately do not necessarily produce accuracy in accord with the precision of the readings. Their powerful scopes, high initial cost, and lack of good compass prevents them from being the best choice for the average rural surveyor.

For practical use, versatility and reliability in rough going, the American light mountain transit with one-minute verniers is hard to beat. The telescopes are seldom over 20x magnification which is about the maximum for woods work. They are light, compact, sturdy and usually have an adequate compass. The age of the instrument is of little consequence so long as it is in good mechanical condition and adjustment. Many of the older transits have compasses of larger size and superior design and construction compared with those now manufactured.

For those who understand its use, the General Land Office pattern solar transits are best of all in the woods. The added cost of the solar attachment will pay for itself in saved crew time very quickly if much transit line is to be run. When using the solar attachment, backsights are eliminated and trees on line are easily bypassed.

The Crew

Normally a 3 or 4-man crew is required for a transit survey because of the need for backsights and additional brush cutting.

The Method

The method recommended is the random traverse—probably the most flexible and economical method for the woodland surveyor. It requires more computations than some other methods but pays off in shortening field time for the crew which results in lower cost to the client. The great advantage to this system is that it allows the surveyor to pick the *easiest* and *most open* route. Although more distance is measured, it allows more rapid progress from point A to point B than if hacking a straight line through briers and brush.

The random traverse is normally run by deflection angles. The difference between the first angle and the mean of the doubled angle should be less than one minute unless extremely steep vertical angles are involved. The mean of the doubled angle is used as the final value. Unless the graduated circle or "limb" of the transit is running eccentrically (determined by tests), only one vernier need be used, *but always the same one.* The deflection angles should be no larger than necessary to provide the best route, avoid trees, brush patches, structures, etc. The flagman or head chainman should always be on the lookout for opportunities to get long sights when selecting the next station.

Field Notes

There are many methods of keeping transit survey field notes. This one is tried and true for reliability and utility. All angular and distance values are recorded in appropriate columns on the lefthand page of a standard field book, *beginning at the bottom of the page.* A running diagram, not to scale, is kept on the righthand page which also shows deflections and distances. It, too, begins at the bottom of the page since in this way the field book is oriented in the direction of progress. This method can save a lot of time and confusion.

A column should be reserved on the lefthand page for observed compass bearings for the foresight at each station. This provides a check on gross blunders and is of great value when making final computations. It also allows a rapid approximation of your actual position at any time by means of latitudes and departures in the BLM Standard Field Tables.

Computations

The random traverse method requires the coordinates of all stations to be obtained by latitudes and departures. To do this, the deflection angles must first be converted to azimuths or preferably bearings. The writer finds it easier to convert the deflections to azimuths and then to bearings. The bearings are then converted to latitudes and departures by multiplying the horizontal distances by the cosines and sines respectively of the bearings. The proper signs (plus or minus) are noted, and obtaining coordinates is merely a matter of addition and subtraction. To avoid negative-value coordinates, begin with coordinates of 10,000 or 100,000 instead of zero. Close work will require the use of 5-place natural function tables. These may be used to the nearest minute for 1:2,000 accuracy, but for 1:3,000 or better, the fraction of the minute must be interpolated, or more complete tables used.

Marking the Line

When the true line is to be marked on the ground, the procedure is this: When computations are completed, the coordinates of all random traverse stations are known. After the bearing of the true lines between corners is computed, calculate the coordinates of points along the true line at the spacing required to provide intervisible points. Ordinarily in the woods this will mean using every random station and sometimes setting several true points from a single random station. The deflection angles and distances from the random traverse stations are precomputed. It is then a matter of returning to the random stations, backsighting on the correct adjacent station, turning off the computed deflection and measuring the calculated distance.

Chaining

So far the discussion has been primarily concerned with angular measurements of lines. A most important aspect of any survey is the linear measurements. Minor angular errors of a compensating nature have little effect on the overall accuracy of the survey. Therefore, the crux of survey accuracy is treated separately because of its extreme importance. The precision of linear measurements must be commensurate with the angular capability of the survey instrument used. If in doubt, always chain with greater precision than you feel would be needed.

Therefore the chainmen should use a good steel tape, and plumb bobs will be required for sighting and measuring. The tape length should be the longest consistent with the topography, vegetative cover and other factors affecting the average length of measurements. It should be the type with an *extra* foot (or link) graduated to hundredths on the opposite side of the zero from the regular graduations.

For closures of about 1:2,000, measure to the nearest 0.10 foot. When going for more precise closures, use the nearest 0.01 foot, which is actually little extra work. Measurements in the woods are nearly always made with the chain suspended rather than being supported along its length. This means that heavy tension must be applied to keep sag from introducing a cumulative error into the measurements. The tension required to get a correct measurement with the tape suspended is a great deal more than is commonly believed. The approximate tension required for true values at 68°F. is listed for various tape types and suspended lengths:

Type of Tape	*100 feet*	*200 feet*	*300 feet*
1/8″ wide	18 lbs.	30 lbs.	37 lbs.
1/4″ wide	26 lbs.	42 lbs.	52 lbs.
5/16″ wide	30 lbs.	50 lbs.	63 lbs.

Unless he is equipped with a tape clamp, the rear chainman should pass the unused part of the tape over his shoulder, across his back, through his crotch, over his leg and then dig his heels in and hang on for long shots. If in bad footing, he may need help in holding a foot mark at the measuring point. The head chainman has a loop thong of convenient length hooked behind his elbow to allow holding the bob string over the graduated extra foot with the other hand while the bob just clears the station mark. He just leans into the loop to get the tension desired after assuring that the tape is clear of obstructions.

The strong tensions recommended may be contrary to some textbook figures, perhaps because most texts appear to assume that the tape will always be supported. The writer has obtained closures of better than 1 in 20,000 with this practice, using a precise transit. The distances have agreed well with controls established by electronic measuring devices which provide very nearly absolute values over long distances. When too *little* tension is applied, trouble develops in using a tape suspended. The chance of straining a good quality modern steel tape to the point of damage is remote if a steady tension is applied.

Specific Methods

Three systems of chaining that work well in the woods are described.

1. *Single-chaining from the instrument.* The rear of the tape is held at the horizontal (telescope) axis of the transit on a foot mark while the head chainman determines the fraction of a foot. Once tension is taken, the rear chainman continually shouts "good, good" so long as the mark is in exact position. The foot is recorded. When the head chainman has observed the fraction, he strikes the tape two quick blows with his hand (or shouts acknowledgement) and *eases* up on the tension. (A sudden release of tension could put both the rear chainman *and the transit on the ground*). The fraction is called out to the notekeeper to complete the recorded slope distance.

The vertical angle to where the head chainman actually held the tape is read on the vertical circle or arc of the transit to the nearest minute. Depending on the accuracy required, the slope distances are multiplied by 3 to 5 places of the cosine of the vertical angle to obtain horizontal distance. Chaining this way can be done to a very high degree of accuracy. However, there are no checks on chaining or angle observation, and locating a blunder may require rerunning the whole traverse. Therefore, great care must be taken with all steps.

2. *Independent chaining.* Each chainman is equipped with a good clinometer (such as an Abney level), a plumb bob and chaining pins. They each keep a separate set of chaining notes showing slope distances, vertical angles to the nearest ¼°, and horizontal distances. The vertical angles should not differ more than ¼° and the recorded slope distances should be identical. These items are checked before starting the next course. Slope distance measurements are made to some multiple of ten units, e.g., 140 ft., 230 links, etc. This will facilitate rapid on-the-spot conversion to horizontal distance *without computation.* The horizontal distance is found in the latitude column of Traverse Tables in the BLM Standard Field Tables.

Before chaining is begun, a visible marker such as a lath, stake, or range pole, is placed on the next transit station. When the distance between stations exceeds the length of the tape, the rear chainman (who is on line) "lines in" the head chainman with the marker for each measurement. When closing on the next station with a slope measurement, the distance is still taken to some multiple of ten units, even though the chaining point falls a few feet ahead of or behind the transit station. If that station is an angle point, it is quicker to measure the difference horizontally and apply the correction, than to convert a fractional slope distance with the tables. The correction is shown in the field notes. Horizontal distances are kept continuously up to date.

This method is well adapted for rapid progress, especially where long spans between instrument stations are possible. The accuracy is adequate where closures of one part in 2,000 to 4,000 are desired. This self-checking system is probably the best choice for most woodland and mountain work because once the target is set, the chaining party and instrumentman are independent of each other. Neither is holding up the other, which leads to covering more distance in a given amount of time than the other methods.

3. *Chaining with High and Lows.* This is a self-checking version of method (1) above, which is used where closures above 1:4,000 are required. Using good tapes and precise transits, second order work has been done with this method. It should therefore serve any need of the rural surveyor. Distances are taken to the nearest 0.01 ft. and vertical angles to the nearest minute. It is described below:

a. The head chainman wraps about three turns of the plumb bob string around his index finger with bob point just clearing the hubtack and his finger as close to the ground as can be viewed by the instrument man.

b. The instrumentman sights the upper side of the chainman's finger, and the vertical angle is recorded as the "low" angle.

c. The distance is measured as in method (1) and recorded as the "low" distance. While doing this, the string is not allowed to slip on the finger, which is held against the under side of the tape while the string is lying over the tape.

d. The whole process is repeated with the head chainman holding his finger (and the tape) about two to three feet higher. The slope distance and vertical angle are recorded as the "high".

e. Before the instrument is picked up, the high and low measurements are reduced to horizontal values, using 5-place cosine tables. This can be done rapidly with a Curta hand-operated calculator. A slide rule can be used with the *versine* tables to obtain the *difference* between slope measure and horizontal distance with almost equal results.

f. Using 5-place cosine tables, the highs and lows should agree within 0.01 ft. for every 100 feet measured. If they do not, something is wrong and they must be repeated until satisfactory agreement is achieved. Give precedence to the "low" when computing the traverse because it is likely to be the more reliable due to less bob swing.

This method may appear to be complicated in writing, but it works smoothly on the ground. While computations are being made, the other crew members can be cutting brush, etc. This method pays for itself by avoiding a return to the scene to locate a chaining blunder, as does method (2).

Costs

Costs of surveys vary widely throughout the country. A review of actual costs of transit surveys versus compass surveys in California forested mountain land shows that the cost of a good transit survey runs about 3 to 5 times that of a good compass survey in the same area. This is caused by a larger crew, slower progress, and more expensive equipment. The biggest expense factor is manpower.

The choice of instrument to use will depend upon land values, the client's desire, and the surveyor's judgment. As a rule of thumb, use a very good compass when closures of under 1:2,000 are in order. (The chaining must be done carefully and sights kept under 200 ft. when 1:1,000 is to be exceeded.) When closures of 1:2,000, or better, are to be attained, use the transit.

REPORT RATING FORM

Name of Report Writer ——————————Name of Reviewer ———————

	WEIGHT	RATING
Title shown (Report- or Certificate-of Survey)	2	
Date of Survey	2	
Name and address of client	5	
What land was surveyed	2	
What records examined	10	
Who was contacted and what information obtained	10	
Problems encountered and how resolved	15	
Encroachments or gaps discussed	8	
Occupancy and possession described	8	
Equipment and methods used	5	
Error of closure of random survey	5	
Name and address of helpers	3	
Disposition of records (plat, description, report)	5	
Typed name of surveyor, and signature	5	
Seal of surveyor	5	
Report written on typewriter	10	
	100	

Chapter 31

SURVEYING CITY RESIDENTIAL LOTS

The author has resurveyed many lots in the city of Elkins and nearby towns. Elkins has about 10,000 people and its survey problems likely are typical of other communities in West Virginia. Elkins originated as a lumbering and coal mining center. Incorporated in 1890 its land was part of the "metes and bounds" surveys of the 13 original colonies. Eventually the town reached a point when there was a need for lot subdivisions; they now cover most of the land in the city.

The Law of Subdivision Markers

The State Code (39-1-17) requires that not less than two "permanent" markers be installed in each "lot or block", and they must be shown on the recorded plat.

The Present Situation and How It Got That Way

Through passage of time, grading and construction activities for buildings, streets, etc., many markers were moved or destroyed. It was customary to set (short-lived) stakes at many lot corners when subdivisions were laid out. Most stakes disappeared—from rot, frost-heaving, bulldozing, or by being pulled up. Owners desiring to build later, often did their own measuring and set stakes or pipes. With inadequate plats and a scarcity of original markers, the replacement of markers at the intended spots was difficult, even for a surveyor. Perhaps not over 50% of Elkins lots now have markers, and some of these are unsuitable stakes; some are in wrong locations.

The intermittent construction of streets and sidewalks added to the uncertainty. There was no fulltime city engineer and not always a county surveyor. Private surveyors were employed to lay out streets and curbs; sometimes State Road engineers assisted. Even now some streets are not curbed or fully paved.

Property owners built fences or walks and planted hedges or shrubs in relation to where they believed their line to be. Some buildings were erected up to the line. A zoning ordinance now prohibits new residential construction closer than five feet to the property line.

The Rules for Lot Surveys

The rules for locating lots are similar to rules for locating rural property. The main difference is that *rural* properties are usually subject to *senior* and *junior* rights, whereas all *lots* in one subdivision normally have *equal* rights, even though all lots are not sold at the same time. The following priorities apply:

1. Natural monuments called for on plat or deed have first priority. Rivers, creeks, and cliffs are natural monuments.

2. Artificial undisturbed monuments or lines established when subdivisions were first laid out have next priority, even though they do not agree with the plat. If a street, for example, is found to measure 50 feet wide between found markers, but the plat shows it to be 40 feet, 50 is likely to control. If a street measures 40 but plat shows 50, it is unlikely that 5 feet can be taken off the lots on each side to add to the street. A major problem is to prove the originality and accuracy of monuments. Old fences between lots may dictate line location. Even erroneous monuments, when long acquiesced in by the public and surveyors, may be presumed to be correct unless proved wrong.

3. If natural or artificial monuments are not present, curbs and pavings of streets adjoining the lots become the best evidence, especially if they agree with the plat, the street and lot lines.

4. Next comes street lines in nearby or adjoining blocks, especially if the plat shows them to be straight. If there are curves, this evidence is more difficult to transfer to another block.

5. Last in priority, plat distances and angles control. Street widths usually take priority over lot widths, because the public interest outranks private interest. If street width is not stated on the plat, the scaled width has weight in absence of other evidence. However, shrinkage and expansion of maps must be allowed for.

Rules for Excess and Deficiency

The distance between markers or streets may not be equal to the footage shown on subdivision plats, because of reasons already stated. Therefor excess (surplus) and deficiency (shortage) may exist. Rules cannot always be found or laid down to determine what to do with excess and deficiency. When doubts cannot be resolved, and values are high, thorough study of texts is required, and possibly an opinion obtained from an attorney. The following guide lines will, however, solve most cases:

1. The chief general rule is to distribute excess or deficit among all lots in a block, or in the area involved, even though they may be irregular in size or shape, *unless there is evidence to the contrary.* Distribution is made in proportion to the relative frontages or dimensions of the lots. *Contrary* evidence may be:

A. If the excess or deficit can be traced to a specific lot, that lot should absorb it. Example: if the last lot or one lot in a tier has no dimension, that lot would usually absorb the difference. If one lot has a different dimension than all others, it would share in the distribution, but this is subject to rebuttal.

B. Original markers of interior lots control.

C. If evidence shows that there should be an unprorated residue, no distribution is made. Example: the plat or resurvey may indicate that the last lot does not reach the subdivision boundary. The larger the gap, the less desirable it is to apportion it. The gap may be owned by original owner or his successors.

D. The lines of possession between lots may ripen into controlling monuments.

E. Small differences—a few inches up to a foot—might better be prorated, whereas large differences often run into barriers of possession evidence.

2. When the plat of a subdivision shows streets or alleys running straight through various blocks, excess or deficit in one block does not often affect other blocks, but should be prorated *within* the block. The streets thus may control the block corners. A "block" usually means the space occupied by lots between streets, even though the plat may show "Block A" encompasses a number of streets.

3. Public streets and alleys ordinarily do not share in excess or deficit, but this may be rebutted with contrary evidence. This could occur if two subdivisions, separated by a street area, were not properly tied together when the later subdivision was laid out. The actual street width could be more or less than anticipated by the subdividers.

4. If a single lot is subdivided by plan or plat, and that lot is entitled to a share of excess or deficit of the entire block, that lot's share must in turn be allotted to the parts of the subdivided lot by proportion.

5. Apportionment of footage will usually result in apportioning square footage or acreage.

6. Value stated in an instrument might affect apportionment. Example: if a grantor divides a lot equally in value but without dimension, it presumes an equal apportionment of excess or deficit.

Procedure for Resurvey of Lot

Practical steps for locating lots and streets are:

1. Abstract the deed of client and other relevant owners.

2. Copy the recorded subdivision plat, noting markers and dimensions. Check unrecorded plats in the City Clerk's Office and Assessor's plat.

3. Look for the nearest original monument in each direction, especially along the frontage line. Excess or deficit is found by measuring distance between monuments and comparing it with the plat dimensions.

4. Make a sketch of lots in the significant area so that actual measurements can be recorded on it. Letter the key points.

5. If there is an authentic marker at each end of the block, measure

the lot front widths with a long tape (300 ft. or more), beginning at one end and working toward the other. Set a spike or nail and cap (N/C) at the plat distance for each lot: show this distance in the notes or on the plat. (Caution: watch for cars running over the tape in driveways.)

6. If there is no marker at block corners, but there is a curbed street at each end: determine center of streets (X) and (Y) (see sketch at left) near (A) and (B) and set pin at edge of each. Do this by measuring between outside of curbs, taking half, and subtracting that from half the street width; measure resulting distance away from curb and set a pin. (Example: if it is 30.1 ft. between curb edges, and street is 50 ft. wide: ½ of 30.1 is 15.05; ½ of 50 is 25. 25 minus 15.05 is 9.95; set pin 9.95 from curb. Allow for crooked curbs.) Do same for street (Z) at (A) and (B). Set N/C or spike at the two intersections of streets by trial and error. Repeat above steps and establish markers at (C) and (D). Markers have now been set at or near the four main block corners.

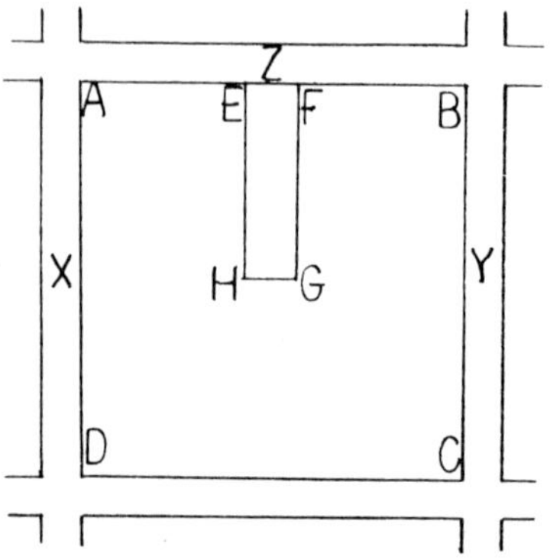

7. Search the vicinity of lot corners for former markers, tie in and show on sketch or notes. Probe below ground with steel pin, use metal detector, or dig if necessary for key markers; remove sod carefully so it can be replaced. Holes left by markers are sometimes evident. Discuss reason for digging with adjoining owners.

8. Measure to, and record location of fences, hedges, alleys, driveways, structures, or other items if they might indicate lot boundaries.

9. Repeat steps 5, 6, 7, and 8 for lots on opposite side of block.

10. Talk to lot owners and neighbors about the existence of markers, when and by whom set. Contact private surveyors. Note: This job might be done in conjunction with item (5).

11. Contact utility company officials—electric, gas, telephone, and water. Utility poles are normally placed on public property near lines, but in Elkins they cannot be used as accurate indicators of line location. Sidewalks are usually on the street at varying distances from the edge. When State Highways go through a city, State plats may be helpful. Street paving is assumed to be in the middle of the public street, but this is subject to proof.

12. Now refer back to items (5) and (6). If no markers are found in the block, and streets are not curbed, the surveyor should search adjoining blocks for markers and street lines.

13. Set up transit at corners of the block containing the lot to be surveyed, and record the corner angles.

14. Now weigh the evidence, relate it to the rules of surveys, and decide where (E) and (F) should be located.

15. Set hub and tack at each front corner of client's lot (E and F). These must be in edge of street (Z) and at selected footage along the front. Use transit to set tacks in line with block corners.

16. Set transit at (E) and (F), turn 90° or the plat angle as influenced by item 13 and by lot markers along (CD). The angles may vary a few minutes from the plat angles.

17. Measure lot depths, considering excess or deficit, if any, and set a N/C at each rear corner (G and H).

18. Consider the original plat distances and angles and adjust (G) and (H) to meet the best compromise of rules of boundary surveys.

19. Reference each final tack by angle and distance (to nearest minute and 0.01 ft.) to the two most permanent objects on the lot or nearby, such as corner of a house foundation, original monuments or other markers. Tie also to curb in front of each front lot corner.

20. Set adequate markers at all corners.

21. If desired, make a closed traverse.

22. Prepare plat showing the new data and references. State area of lot to nearest 5 square feet.

23. Make a Certificate or Report of Survey to show the reasons for the decisions made.

24. Keep the adjoiners informed of progress to the extent necessary. A copy of the plat and certificate may be furnished to them.

25. If evidence heretofore described is missing, the surveyor has little to guide him except the plat and his judgement. The intent of the subdivider as shown by the plat, should be reflected to the maximum extent practicable in the decisions of the surveyor.

Miscellaneous Items

1. It is better to start to measure from interior streets than from a property boundary, because the latter is less certain.

2. It is usually best to set markers flush with ground. Guard stakes should be set if construction equipment is to be used nearby.

3. If a surveyor wants a certain lot of a new subdivision to absorb the excess or deficit, he should not show any dimension on the plat for that lot. The Report of Survey should mention this.

4. If parts of a non-subdivided tract are sold by one grantor at different times, the last grantee is the loser or gainer.

A Word to the Property Owner

After reading this chapter, a lot owner will know some of the difficulties in relocating his property markers. This knowledge, however, contains pitfalls for the person who concludes that he can set his own markers without consulting a surveyor. He can, of course, measure for his own information. But marker location affects neighbors and the public street. If a dispute develops, the owner cannot expect his efforts to be cordially received by his neighbors, nor as technically and legally satisfactory by a judge and jury. When a witness has an interest in property contested for, his testimony may be prejudiced or inadmissible. The *wise owner* will employ a competent surveyor, first for advice, and then to resurvey the lot.

A client may assist in measurements. But if there is a known dispute over a line or corner, it may be better for the client not to take an active part in the work.

A Word to the Surveyor

Surveyors face a challenge to help establish an adequate number of properly-set markers in each block. Their location must be made known to other surveyors and the public by urging the client or purchaser to record the plat. Because of the danger of disturbance of markers and reference points, three or four markers per block would be good, some being at the block corners. If the client's lot is not on the corner, the lot corners, well referenced, will make a good beginning.

If natural and artificial original monuments are scarce, as they often are, the curbs and pavings of streets are readily available objects of semi-permanent nature. A hedge or fence, though originally on line, can be removed and cannot be relied upon. The value of careful surveys by towns and cities for new curbs and pavings can hardly be overestimated.

Chapter 32

TITLE INSURANCE SURVEYS

Title insurance involving a transfer of title to real estate consists of two parts:

1. Matters pertaining to the quality of the title
2. Matters pertaining to the location of the property

If the insurance desired relates only to the quality of the title, the attorney checks only matters of title, and rarely things on the ground. If, however, the title must be insured "without exception", a surveyor must be employed to certify regarding ground conditions.

There are many title insurance companies in the United States. Among the larger are Lawyers Title Insurance Company, Louisville Title Insurance Company, and American Title Association. The ATA and the American Congress on surveying and Mapping have agreed on *Minimum Standard Detail Requirements for Land Title Surveys.* A summary of these requirements follows:

1. Plat to contain: name, address and signature of licensed surveyor who made survey, his official seal and license number, date of survey, and caption *Land Title Survey* with this certificate: "To (name of client) and (name of title insurance company): This is to certify that this plat and the survey on which it is based were made in accordance with the 'Minimum Standard Detail Requirements for Land Title Surveys' jointly established and adopted by ATA and ACSM in 1962. (signed) ————(SEAL) License No. ————."

2. Title insurance company or client, when ordering survey, should notify surveyor that a *Land Title Survey* is required, and furnish record description, easements, covenants or servitudes affecting property, names and deed data of all adjoiners, and all other pertinent information. Surveyor should be advised if area, zoning classification, or setback lines are to be shown, also of additional Federal Housing Administration or Veterans Administration requirements.

3. The allowable positional tolerances of corners may not exceed:

(a) 0.02 ft. in urban blocks where buildings can be erected along the property line, or where high land values warrant it.

(b) 0.04 ft. in urban or suburban subdivision interior blocks and/or urban and suburban lots or parcels.

(c) 1.0 ft. per 5000 ft. of perimeter in rural areas, except:

(I) Closer tolerance required if values are increased by nearness to major highway intersections or complexes, building congestion, oil or mineral rights, etc.

(II) When the land is extremely long or narrow, closer tolerance is required on the shorter narrow dimensions to qualify acceptable corner positioning in relation to the narrow width.

Surveyor shall note on plat: "Maximum positional tolerance of corners is ———."

When surveyor has doubt as to ground location of street or lot lines being within above tolerances (because of indefiniteness, insufficient markers, or errors in descriptions or plats), surveyor shall indicate nature of difficulty and give opinion as to range, scope and effect of same on the surveyed positions.

4. Scale of plat should be shown and be convenient, with graphic equivalent. North should point to top of drawing if practicable. Supplementary or exaggerated-scale diagrams should be to scale. Minimum drawing size 8½″ x 11″.

5. Plat shall contain the following applicable information:

(a) All data needed to show the mathematical dimensions and relationships of the boundary, with closure not less than 1:10000; with angles given directly or by bearings; curve length, radius, chord, and chord bearings. Bearings should refer to true north, State Plane Coordinate north or some well-fixed bearing line so that bearings may easily be reestablished. North arrow, preferably in upper right, must be referenced to its bearing base. Boundary bearings should be clockwise.

(b) When record data differ from measured data, both recorded and measured data (bearings, angles, distances) shall be shown.

(c) Measured and recorded distances from corners of surveyed parcel to nearest r/w lines or streets in urban or suburban areas and to "found" lot corners. If needed, distance to intersecting streets in both directions from the parcel, also bearings of streets, their names, legal lines and widths. Note changes in street lines or roads, with date of change and authority under which made.

(d) Identifying title of recorded or filed maps pertaining to the parcel, with filing dates and map numbers. Lot, block and section data. Names of adjoining owners and/or recorded parcel numbers and similar data. Interior parcel lines to show contiguity, gores, and/or overlaps.

(e) Monuments, stakes or marks found or placed (distinguish which is which), on premises (or off, if used to establish corners of premises). Character and location of possessory evidence; if none on any line, state "No physical evidence of line."

(f) Kind, character and distance from boundaries of buildings on parcel. Street number of parcel. Physical or known evidence of easements including roads, r/w, watercourses, drains, telephone, telegraph or electric lines; water, sewer, oil or gas pipelines, etc., on or across premises, and on land adjoining if affecting the premises surveyed; surface indications of underground easements. If no buildings on premises, state "No buildings." Kind and location of walls (independent, division, party or other) and whether or not same are plumb. Location of both sides of party walls and thickness. Buildings or fences within two feet of either side of boundary lines. Note if buildings on premises use wall of adjoining property or vice versa. Evidence of encroaching structural appurtenances and projections, such as fire escapes, windows opening out, flue pipes, stoops, eaves, cornices, area ways, steps, trim, etc., *by* or *on* adjoining property or streets, and extent of such encroachment. Openings such as windows and doors in walls of adjoining premises within two feet of boundary lines of premises. Client or insurance company to obtain permission to enter upon adjoining land as needed, otherwise surveyor is not obligated to enter. Location and width of joint, common or independent driveways, ribbons and aprons. Cemeteries and burial grounds. Springs, streams, rivers, ponds, lakes bordering on or running through premises. Streets not physically open.

6. As minimum, surveyor should furnish two sets of plat prints to company or client. Prints should be on durable and dimensionally stable material of quality acceptable to company. Furnish at least two copies of boundary descriptions, which should include date of survey, surveyor's file number if any, political subdivision, etc.

Note by author: Show these items on plats in descriptions and certificates:

State, county, district, city, town

Latest deed or will book and page number, and or previous title source.

Amount and direction of declination

If there are no encroachments, this statement may be used on plats and in supplemental certificates: "There are no encroachments on or by the premises surveyed."

This Is Land Surveying

Land Surveying is 50% retracement, 50% law, and the remainder measurement.—From Vermont CORNERPOST

NOTE: See American Congress on Surveying and Mapping Journal, December 1973, page 536, for Committee Report.

Chapter 33

UNDERGROUND COAL MINE MAPS IN WEST VIRGINIA

The West Virginia Code (22-2-2) of 1969 allows licensed land surveyors to map underground coal mines. The surveyor's seal, signature and certificate shall appear on said maps. The scale of maps shall be not less than 100 feet per inch nor more than 500 feet per inch. The following numbered map contents are quoted verbatim from the law:

1. The shafts, slopes, drifts, tunnels, entries, rooms, crosscuts and all other excavations;
2. As far as practicable the outline of existing and extracted pillars shall be designated;
3. The direction of all air currents, using arrows;
4. The abandoned portion or portions of the mine;
5. The outcrop of the coal bed within the bounds of the property assigned to the mind;
6. The boundary lines of the coal rights assigned to the mine;
7. The known underground workings in the same coal bed on the adjoining properties within one thousand feet of such mine workings and projections;
8. The elevation of the top and bottom of each shaft and slope, all drifts and the bottom of the coal along the haulage entries in each set of main and panel entries at horizontal intervals not exceeding two hundred feet, with contour lines at not more than twenty feet vertical intervals;
9. The location of the principal streams and bodies of water on the surface;
10. The location of any impounded bodies of water inside mine;
11. The location of all boreholes penetrating the coal bed mined;
12. The location of all oil and gas wells, high pressure pipe lines, high voltage power lines and principal roads; and
13. Where the overburden is less than one hundred feet, occupied dwellings shall be designated.

Surveying calculations and mapping of underground coal mines which are opened or reopened shall be done by the rectangular coordinate traversing method and meridians carried through and tied between at least two parallel entries of each development panel and panels or workings adjacent to mine boundaries or abandoned workings. These surveys shall originate from at least three permanent survey monuments on the surface of the mine property. The monuments shall be clearly referenced and described in the coal mine operator's records. Elevations shall be tied to either the United States Geological Survey or the United

States Coast and Geodetic Survey bench mark system, be clearly referenced and described on such map.

Underground coal mines operating on the effective date of this section, and not using the rectangular coordinate traversing method shall, within two years of the effect date (July 1, 1969), convert to this procedure for surveying calculations and mapping.

A diligent search must be made for all existing and available maps and survey data for the workings on the adjoining properties. Proof must be offered that a suitable method was used to insure accuracy in the methods used in transposing other workings to the map of such mine.

When any underground coal mine is worked out, abandoned or closed indefinitely, the operator of such mine shall, within fifteen days after the cessation of the production of coal, have completed, or caused to have completed, a final survey of such mine. Not longer than thirty days after coal production ceased, the operator shall have extended, or cause to have extended, the map of such mine to accurately show all excavations in such mine and a true copy of such mine map sent, by certified mail, to the archive of underground coal mine maps, Office of the Director of the Department of Mines, State Capitol, Charleston. Final coal mine maps shall have thereon descriptions of all survey monuments.

Any person having a map or surveying data of any worked out or abandoned underground coal mine shall make such map or data available to the Department of Mines to copy or reproduce such material.

Any person who fails or refuses to discharge any duty imposed upon him by this section shall be guilty of a misdemeanor, and, upon conviction thereof, shall be punished by a fine of not less than five hundred dollars nor more than one thousand dollars.

The surveyor's certificate shall be in the following form:

> "I, the undersigned, hereby certify that this map is correct and shows all the information, to the best of my knowledge and belief, required by the laws of this state, and covers the period ending ————————."

____

Land Surveyor

Chapter 34

TREE IDENTIFICATION FOR SURVEYORS

Deeds frequently call for trees as corners, and surveyors need to use trees for monuments. It is therefore necessary for the surveyor to know how to identify trees, either the exact species or the group to which it belongs.

Early surveyors sometimes merely called for a birch, without specifying the kind. Likewise they did not try to determine the kind of ash, aspen, elm, hickory, or magnolia. Some species might or might not be differentiated: sugar and red maple, oaks and pines. Pitch and shortleaf pine were more often called yellow or hard pine. Hemlock was known as spruce-pine and basswood was lynn or linn. Sugar maple was often called hard maple or just "sugar"; red maple was soft maple or just maple. One corner was described as a "sugar and a maple", although both are maples. Fire cherry though not often used as a corner might be called "roundwood". Men on the survey crew or oldtimers in the area are good sources of information on unusual names.

The modern surveyor should try to describe the exact species when he uses a tree for a corner or reference; this increases the chance of finding the correct evidence in the future. With study and experience the surveyor should be right 95% of the time. Sometimes a glance at the bark or other things about a tree will identify it; again one may have to examine leaves, twigs, flowers, or fruit; or the clue may be exposure, elevation, soil, or the section of the country. If absolute identity is needed, a forestry expert can be consulted.

Nearly all State Forestry Departments publish booklets which describe common trees found in forests of the state. Other sources of tree identity are given at the end of this Chapter. The surveyor will do well to acquire one or two good publications and to know where others can be found—usually your nearest forester's office.

The following data is given mainly for the Appalachian surveyor who has at least a general knowledge of trees, and who desires to identify most of the common trees with a reasonable degree of accuracy. He will likely need to study other texts having pictures or drawings and more complete data. This chapter discusses features that distinguish a tree from its "look alikes". Although the bark of many trees becomes easily recognizable with practice, bark is difficult to describe and often differs between young and old trees, so only unusual bark characteristics are listed.

Trees are generally placed in two groups: those that lose their leaves each winter (deciduous—sometimes called "hardwoods" although the wood of some is soft) and those that retain their leaves or needles for long periods (evergreens—sometimes called "softwoods" although the wood of some is hard).

Evergreens

Listed in general order of their occurrence in West Virginia are: pines, hemlock, cedar, and spruce.

White pine. Always five needles in cluster; cone 4-8″ long, ¼ as wide; one branch-whorl growth each year.

Shortleaf pine. 2 and 3 needles per cluster; cones 1½-2″ long; young twigs round and purplish; bark has resin ducts.

Pitch pine. Nearly always three needles in cluster; cones 2-3″ long; young twigs angular and brownish; bark has no resin ducts.

Scrub or Virginia pine. 2 shorter twisted needles per cluster; cone about 2″; branches tend to be crooked and persistent. Purplish waxy bloom on twigs.

Table mountain and *red pine.* Not common enough to be described. Scotch pine not a native tree.

Hemlock. Needles 0.3-0.7″ long, flat, blunt-tipped, dark green above and whitish below, in flat row on each side of twig; cone ¾″ long; damp locations.

Cedar. Red cedar (juniper); two kinds of needles: one is flat and scaly, the other sharp, resembling the pfitzer juniper (a common low spreading ornamental). White cedar (arborvitae) has only flat needles.

Spruce. Found near top of highest mountains; needles ½″ long, 4-sided and pointed, growing out from all sides of twig; cones 1-2″.

Deciduous

These are listed below in groups according to their general frequency of occurrence. Some listed as rare are common in specific areas, such as elm along watercourses.

Most Common	*Frequent*	*Less Frequent*	*Rare*
oak	gum	cherry	blue beech
beech	chestnut	magnolia	elm
hickory	birch	cucumber	buckeye
maple	basswood	ash	willow
poplar	sassafras	sycamore	mulberry
locust	sourwood	walnut	aspen
dogwood	sarvis	hawthorn	ironwood

Deciduous trees can be divided into two groups; those with opposite leaves and branching, and those with alternate. This knowledge is helpful when identifying trees by looking up into their tops. There are few species in the "opposite" group, and they are quite easy to distinguish as the leaves of each are different.

Opposite.

Ash. Leaves opposite, but divided into 7-11 leaflets ("compound" leaves); they resemble hickory or walnut leaves until the opposite branching is noted; bark may look like pignut hickory but may have large areas flaked with white mossy material.

Dogwood. Has a plain-type leaf; bark broken into small squarish blocks resembling black gum and persimmon; low and spreading.

Sugar maple and red maple. The two most common maples and often difficult to tell apart in winter. Leaves about same size and general shape, but the edges of sugar are smooth whereas edges of red maple are serrated like teeth of a saw. Bark of young trees looks alike but as trees get large, red has more shaggy look. Sugar buds are long, sharp and brown, while red buds are short, round and reddish. Examine leaves on the ground.

Buckeye. 5-7 leaflets grow out from one common point instead of along the leaf stem; usually found along streams.

Alternate.

Perhaps the best way to start on this large group is with those with the more distinctive leaves. Locust, hickory and walnut are the only species with alternate compound leaves.

Black locust. Leaflets are small—1-1½″ long; two sharp spines grow from the base of every leaf. Honey locust is not common; it has forked spines 1-2″ long.

Walnut. Leaves resemble ash but are usually twice as long. White walnut (butternut) has gray bark whereas black walnut bark is dark, thicker and deeper ridges. Twig pith is chambered.

Hickory. Several species; usually 5-7 leaflets; the most common being pignut and shellbark (with scaly bark). Pignut is much more common. The surveyor will seldom need to tell the exact species.

Sycamore. Leaf and bark like no other; occurs almost entirely along streams or near watercourses or hillside drains.

Poplar. Also called yellow poplar or tulip; the only tree with a squarish end to the leaf; bark crevices in young trees are usually whitish; this color disappears later. One of straightest trees.

Sassafras. Leaves have 3, 2, or 1 lobe; twigs are bright glossy green with distinctive taste and odor.

Mulberry. Some leaves have one thumblike lobe but never three.

Willow. Leaves 3-6″ long but only about ½″ wide; none like them; grows along streams.

Aspen. Two species, much alike except for size of leaf. Bark of both trees is whitish green with raised dark warty bands. The trunk of the tree tends to be straight. Unnecessary to distinguish between species.

Magnolia. Often called wahoo or umbrella tree; several species have largest leaves of any tree—10 to 24″ long; bark smooth and gray.

Cucumber. A magnolia, but with ordinary shaped leaves 6-10″ long; bark something like ash but darker; peculiar fruit.

Oaks. About 14 species of oak grow in West Virginia but only 6 are common—white, chestnut, red, scarlet, pin, and black—these the surveyor should learn. *White oak* is easy to tell by its rounded leaf lobes and thin-looking gray sheddy bark. *Chestnut oak* (rock oak) leaves do not have the usual oak leaves but look more like fat chestnut leaves with wavy edges; its bark is gray with thick distinctive ridges. *Scarlet and pin oak* have the deepest cut leaf lobes—almost to the midrib. Scarlet oak tends to grow straight with side branches persisting. *Pin oak* has leaves like scarlet and tends to grow straight, but its lower branches tend to droop more; is confused with scarlet, but bark is not as rough and acorns only ½″ wide (half the size of scarlet's fruit). *Red oak* and *black oak* may be confused, but leaves of red are more symmetrical and not so deeply cut. Inner bark of red is reddish while that of black oak is yellow.

Gum. Sweet gum has a star-shaped leaf and is found only along a few rivers. *Black gum* is common over the state; has an ordinary leaf 2-3″ long that is one of the first to turn red in fall; bark is gray and on older trees is in squarish chunks; branches tend to a horizontal appearance. The tree is identifiable from a long distance.

Chestnut. Blight kills all trees that live to 6″ dbh. Large and small dead trees are common, unless they have been cut for rails or posts; usually grow in clumps; leaf 5-10″ long with coarse teeth. Stumps long-lasting.

Basswood. Heartshaped leaf with one side larger than the other; ordinary bark; grows well on north slopes, often in clumps.

Hawthorne. Can be told by its small leaves (1-3″) and thorns 1½-2½″ long; small flattopped tree that often grows in thickets.

Sourwood. Leaves 5-7″ long and not wide; a small tree, bark resembles sassafras or elm; twigs are sturdy and reddish; the long clustered fruit of small capsules often persists into the winter.

Leaf Lookalikes

From a distance, the leaves of the following trees look somewhat alike: beech, birch, cherry, ironwood, sarvis, elm, and blue beech. Leaves are 2-5″ long, ordinary looking and with serrated edges.

Beech. Best known by its smooth gray bark and long sharp buds; leaves persist into the winter.

Cherry. Two species—*black* and *fire cherry.* The bark gives these trees away. Black or wild cherry has a dark bark that breaks into squarish chips; twigs and inner bark have bitter almond taste and odor.

Fire cherry grows on old burned areas; bark is smooth, dark with loose papery strips that pull off around the tree; seldom lives long or gets over 10″ dbh.

Birch. Three species—*black, yellow* and *red* in that order of frequency. All have very fine twigs. *Red birch* grows only along river banks at low elevation; bark is whitish to reddish and peels off in papery layers which make a ragged appearance. *Black birch* bark on young trees is dark and smooth with horizontal lenticels like fire cherry; on older trees the bark looks like black cherry but is more irregular; twigs and bark have a sweet taste and odor. *Yellow birch* bark soon becomes silvery or yellowish gray and looks ragged.

Elm. American and *slippery elm.* Most common along watercourses; leaves thick and rough below (slippery is rough on top, too); bark is ordinary looking, like basswood, sourwood, and sassafras. One of the more difficult trees to identify, maybe due to its infrequency.

Blue beech. Also called water beech, as it grows along streams; seldom over 6-8″ dbh; has smooth blue-gray fluted bark; easy to tell.

Ironwood. Seldom larger than 10″ dbh; bark fine and scaly but ordinary looking like sarvis; leaf looks like birch.

Service berry. Or sarvis, shadbush, juneberry, etc. Seldom bigger than 10″ dbh; first tree to flower in spring (white, small, and delicate); July fruit is sweet and edible; bark like ironwood but smoother.

Miscellaneous. There are other uncommon species that grow in Appalachia, but they are not described here. For example, carolina poplar, hackberry, pawpaw, crabapple. Holly is more frequent; its spiny "evergreen" leaves can hardly be mistaken.

Reference Material

Manual of the Trees of North America by Sargent. Houghton Mifflin Company, Boston, Mass., or Dover Publications, Inc., 180 Varick St., New York, N. Y. Complete and scientific.

Knowing Your Trees by Collingwood and Brush. 1947. American Forestry Association, Washington, D. C. Contains excellent pictures and descriptions of trees and woods.

1949 Yearbook of U. S. Department of Agriculture: Trees. From Superintendent of Documents, U. S. Government Printing Office, Washington, D. C. 20402. Good descriptions and a key to identification (p. 833). A good book for the whole family.

The Wood Handbook. U. S. Department of Agriculture Handbook No. 72. 1955. Superintendent of Documents, U. S. Government Printing Office, Washington, D. C. 20402. Technical.

Forestry Handbook. 1956. The Ronald Press Company, New York. $15.00. Contains an identification key for trees and woods. Indispensible to foresters and related professions.

Common Forest Trees of West Virginia. 1968. Forestry Division, Department of Natural Resources, State Capitol, Charleston, West Virginia 25305. 50¢.

Illustrated Guide to Trees and Shrubs by A. H. Graves. 1955. Harper and Brothers, New York. Woody plants of northeastern United States and adjacent regions.

Best Trees for Corners or References

Long-lived species that will tolerate shade are best: Beech, hemlock, and sugar maple. Close behind are: spruce, dogwood, black gum, followed by basswood, cucumber, white oak, poplar, and elm. The ideal tree is 4-8″ dbh, hopefully *too crooked for lumber*, but likely to survive among its neighbors.

And You Think You Have Red Tape!!!!

A New Orleans lawyer sought an RFC loan for a client. He was told that the loan would be granted if he could prove satisfactory title to property offered as collateral. The title dated back to 1803, and he had to spend three months running it down. After sending the information to RFC he got this reply: "We received your letter today enclosing application for loan for your client, supported by abstract of title. Let us compliment you on the able manner in which you prepared and presented the application. However, you have not cleared the title before the year 1802, and therefore, before final approval can be accorded the application, it will be necessary that the title be cleared back to that year."

Annoyed, the lawyer replied, "Your letter regarding titles in case #18915 received. I note that you wish titles extended further back than I have presented them. I was unaware that any educated man in the world failed to know that Louisiana was purchased from France in 1803. The title to the land was acquired by France by right of conquest from Spain. The land came into possession of Spain by right of discovery made in 1492 by a sailor named Christopher Columbus, who had been granted the privilege of seeking a new route to India by the then reigning monarch, Isabella. The good queen, being a pious woman and careful about titles, almost, I might say, as the RFC, took the precaution of securing the blessing of the Pope for the voyage before she sold her jewels to help Columbus. Now the Pope, as you know, is the emissary of Jesus Christ, The Son of God, and God, who, it is commonly accepted, made the world. Therefore, I believe it is safe to presume that He also made that part of the U.S. called Louisiana, and I hope to hell you are satisfied."—WEST VIRGINIA SURVEYOR

Chapter 35

WOOD IDENTIFICATION FOR SURVEYORS

The surveyor's need to identify the species of wood is less frequent than telling the kind of tree. But on occasion the species of a stump, log or root can be important. It would not be proper to decide that a certain stump was the corner unless it was the species called for (or was a very similar species).

The trees most commonly called for in West Virginia deeds are: oak, chestnut, hemlock, beech, hickory, maple, poplar, locust, gum, birch, dogwood, sassafras, cherry, cedar, lynn, pine, spruce, ash, cucumber, sycamore, sarvis, walnut, and elm. This group deserves the most study for wood identification.

For wood identification the surveyor should acquire four things: a good key scheme, a collection of wood samples, a very sharp knife, and a hand lens of 2 and 4 power. The USDA *Yearbook Trees* and the *Forestry Handbook* have good and identical keys. Specimens in important cases may be sent for identification to the U. S. Forest Products Laboratory, P. O. Box 5130, Madison, Wisconsin 53705.

The surveyor should make his own collection from logs, stumps, roots, sawmills, wood processing plants, etc. All should be carefully labelled as to species and, if on a survey, where found.

To examine a wood specimen, try to obtain a piece of heartwood, because (a) the keys are based on heartwood, and (b) the sapwood is often thin and rotted away. Then saw or cut so as to obtain an end (cross section) and two longitudinal (side) sections at right angles to each other (a radial section parallel to the tree radius, and tangential at right angle to the radius). To understand wood technology, a surveyor should learn some terms:

1. *Pores.* Rounded openings on end cuts, and fine grooves on side cuts. Pores may be so small that they are not visible to the naked eye, or they may be large enough to be easily seen.
2. *Ring-porous.* When large spring pores are visible to the naked eye and the small summer pores are not.
3. *Diffuse-porous.* No large pores; small ones may be a little larger on the spring growth.
4. *Rays.* Marks, lines or bands differing in color from surrounding wood and containing no visible pores. Rays visible on end and side cuts.
5. *Annual rings.* The wood grown in one year.

The following key is adapted from the key in USDA *Yearbook Trees,* with comments by the author. Only the more common species are listed. The key is based on heartwood (usually darker than sapwood) unless otherwise stated. It is based on fresh-cut wood with average ring width, and *without a lens.* A lens, however, may clarify minute details.

HARDWOODS

I. Pores visible.

A. Ring-porous.

1. Summerwood on end cut has light and dark irregular V-radial designs or patches.
   a. Many rays broad and conspicuous; wood heavy. *Oak.* Two wood groups—white and red; it is possible only to tell which group the wood belongs to.
      a-1. Wood has reddish tinge; large pores usually open. *Red Oaks* (red, black, scarlet, pin, spanish).
      a-2. Wood usually yellowish brownish without reddish tinge; large pores mostly closed (except chestnut oak). *White Oaks* (white, post, chestnut).
   b. Rays not noticeable; color brownish; wood lighter than oak. *Chestnut.*

2. Summerwood has short or long wavy tangential lines.
   a. Heartwood brown with red tinge.
      a-1. Tangential bands long, conspicuous in summerwood: *Elm.*
      a-2. Bands short, inconspicuous and in outer summerwood; springwood several pores wide. *White Ash.*
   b. Heart yellowish to greenish brown; pores closed; rays indistinct; one of our heaviest woods. *Black Locust.*
   c. Heart silvery brown, with spicy odor and taste; rays indistinct; wood heavy. *Sassafras.*

3. Summerwood radial or tangential patterns not visible without lens; heart reddish brown; sap over 1″ wide; one of our heaviest woods. *Hickory.*

B. Diffuse-porous.

1. Single pores plainly visible on end and side, larger in springwood; rays smaller than largest pores; pores not in radial groups across rings.
   a. Heart chocolate or purplish brown; wood heavy. *Black Walnut.*
   b. Heart light chestnut brown with springwood darker than summerwood. *White Walnut.*

2. Single pores barely visible in good light and smooth surface;

pores fairly uniform size throughout, not crowded on end cut; heart red-brown; wood heavy. *Birch.*

II. Pores not visible.

A. Rays broad, conspicuous; heart shades of light reddish brown.

1. Rays crowded on end, up to 3/16″ long on side cuts, make pronounced crowded "flakes" when quartersawed; summerwood not darker or denser; interlocked grain; heavy. *Sycamore.*

2. Broad rays not crowded, up to 1/8″ long on side cuts, producing scattered "flakes" when quartersawed; distinct, denser darker band of summerwood; straight grained and heavier than sycamore. *Beech.*

B. Rays not conspicuous but visible.

1. Heart deep rich red-brown; annual rings well defined; rays distinct. *Black Cherry.*

2. Heart light grayish brown with red tinge; rings clearly defined by thin darker reddish-brown layer; rays very distinct.
   a. Wood heavy and hard to cut across grain; pith flecks rare. *Sugar Maple.*
   b. Wood not so heavy and easier to cut across grain; pith flecks often abundant. *Red* and *Silver Maple.*

3. Heart light yellow-brown, with greenish tinge, sometimes purplish; sap wider than 1″; rings and rays distinct.
   a. Wood moderately light. *Yellow Poplar.*
   b. Wood moderately heavy. *Cucumber.*

4. Heart creamy brown; sap wide, not sharply defined from heart; rays fairly distinct; very light wood. *Basswood.*

C. Rays not distinctly visible

1. Rings not clearly divided into band of soft springwood and denser and darker band of summerwood; rings inconspicuous.
   a. Heart distinctly darker than sap.
      a-1. Heart reddish-brown; wood fairly heavy. *Birch.*
      a-2. Heart grayish brown; wood crossgrained. *Black Gum.*

SOFTWOODS

2. Rings clearly divided into band of soft springwood and denser darker band of summerwood; rings more pronounced than 1.
   a. Wood resinous as shown by exuding pitch, pitch streaks

and pockets especially if heated, or by longitudinal resin duct lines from less than 1 inch to several inches long.

a-1. Heart darker than sap, reddish brown or orange brown; resin ducts abundant.

b-1. Summerwood not conspicuous and not much darker or harder than springwood. Wood light or moderately so. *White Pine.*

b-2. Summerwood conspicuously darker and harder than springwood; wood heavier than b-1. *Other Pines.*

a-2. Heart not much darker than sap; resin ducts scarce and inconspicuous; wood has no distinct odor. *Spruce.*

b. Wood not resinous.

a-1. Heart darker than sap, light brown to dark reddish-brown; aromatic odor and taste. *Cedar.*

a-2. Heart not much darker than sap when dry; all wood pale reddish brown. *Hemlock.*

Dogwood is not listed in the key. Rings are present; pores numerous, small to minute but distinct with lens, well distributed and rarely radially connected; rays heterogenous of two sizes; heart small and red brown to light chocolate; sap pinkish; wood hard and our second heaviest.

Sarvis is not listed in the key. Thick brownish sapwood; the brown or red brown heart is absent on small trees; brown pith flecks; wood hard and probably our heaviest; grain straight.

Durability of Wood

The surveyor may occasionally need to set a wood marker at a corner or for reference, although wood is inferior to stone or iron. The *heartwood* of certain species such as black locust may last for 30 years or more. The following list of the more durable species (heartwood) is in order of durability in contact with the ground, as stated in the *Wood Handbook:* cedar, chestnut, locust, mulberry, and black walnut. Cypress, catalpa, and osage orange are good if they can be obtained. Sarvis heartwood is said to be durable. Moderately durable are: white oak, sassafras, and white pine.

Chapter 36

MISCELLANEOUS PROCEDURES AND TECHNIQUES

The One-Man Survey Crew

Many surveyors in West Virginia—and perhaps in other states—are "small operators". Occasionally a job must be done without any help; sometimes it is the least costly method.

Rural Surveys

One-man work is usually limited to corner searches, etc. But it is possible for one person to establish a straight line through fields or woods by leaving markers at intervals and lining up the next one by eye. If the line misses the desired objective, another random line can be tried. Distances can be paced or estimated.

City Surveys

One man can sometimes survey a city lot. He can often set markers as easily as he can explain to a helper what to do.

Distances. The measurement of distance with tape can be done thus: Tie a leather thong at the zero end and tie two or three loops in it. Set a pin firmly in the ground in each loop so when the proper tension is applied the zero mark of the tape will be over the hub tack. If the distance to the other hub is not an even foot, set a pin at the nearest foot and measure the additional foot-fraction with pocket tape or rule, and record the fraction. To increase accuracy, both ends of the tape may be pinned down; this allows the zero end to be checked under tension. If ground slopes, use a plumb bob at the lower end.

Angles. Set tacks part way in so they can be seen through the telescope. Or set a nail in the hub in line, or set a target and plumb bob over the tack. A target tripod can be made for a few dollars thus: Buy three 8′ aluminum rods about 3/8″ thick; drill a 1/8″ hole through each about ¼″ from one end. Out of a 1″ piece of hardwood about 3″ square, fashion a head as shown in the plan to the left; drill holes for bolts and fasten a rod to each bolt. Sharpen the bottom of each rod. Place a screwhook for plumb bob under the head. Four tripods can be used effectively. Set one over each marker, moving it aside when the transit is to occupy the corner. To adjust: With fingers on bottom end of rod, and hand or wrist on ground, push up or down gently.

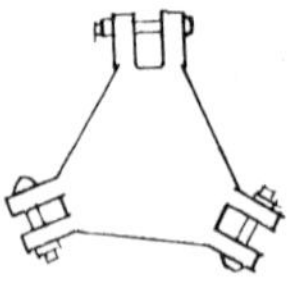

Pacing

Approximate distances may be measured by counting paces (2 steps per pace). Pacing should be done at the *normal gait,* counting every other step. By practice, one's normal pace on smooth level ground can be determined within 0.1 ft. For many people 5.2 to 5.3 is normal. Do not try to take a 6 ft. pace unless it is normal for you. As ground becomes rougher or steeper, the pace shortens. The number of paces can be kept straight by clenching one finger in the fist for each 100 paces. Practice pacing on 2-ch. survey shots; it helps locate obscure stakes or marks.

The Picket Line

A picket line is one established by lining up stakes by eye. A compass may be used for the starting direction. A 3-man crew is best. Stake #1 is placed at the beginning point; #3 is placed on line of sight about as far as can be easily seen. #2 is placed in between. The crew leader sets future pickets. #4 is set in line with #2 and #3; #5 is set beyond #4, in line with as many pickets as can be seen; the more the better. Space pickets about 1 ch. apart. The helpers clear brush and make pickets. Repeat the process until closing corner is reached.

Picket stakes should be 3 to 5 ft. long, straight, with a colored ribbon tied in a neat bow at the top. Set pickets so the *ribbons* are on line. Ribbons may be tied on solid branches or saplings or on small aluminum nails driven into tree bark. Since spacing is not exact, pickets can be shifted to soft ground.

When a line hits a hollow, the leader suspends a plumb line so ribbons on the opposite slope are in line with the string. When a line crosses a ridge, a plumb line or short pickets can be used. To get by obstacles, the last 2 or 3 pickets can be offset, then extend the offset line, and offset back again. If the obstacle is a small tree, the leader passes the tree to a normal distance and sights back *alongside* the tree to the back line, then along the other side of the tree, and sets the picket midway between these two lines of sight. The leader moves on, repeats the process until the obstacle is well behind.

When a point at right angle to the closing corner is reached, measure offset, divide by length of the line and get the offset per chain or 100 ft. Measure back roughly along the line a given distance, measure the offset, and set a stake. Pickets not needed should be pulled up and cast aside.

Making a Right Angle With Tape

A right angle can be turned accurately with a tape. The method is based on the theory that the length of the sides of all right triangles are in proportion of 3, 4, and 5. If the short side is 6 ft. long, the hypotenuse will be 10 ft. and the remaining side 8 ft. One man can lay out the triangle by setting pins at thc ends of the "4" side, fastening the end of a tape at each end and intersecting the "3" and "5" marks of the two tapes. The longer the sides, the more accurate the triangle. A problem in using this method in woods surveys is to know if the "4" side is located on the true survey line.

Old Marks

Looking for Old Marks. Previous survey marks are often confused with marks made by timber cutters and others. Two or 3 hacks or blazes, or blazes with hack above or below, are more likely to be survey marks, especially if several are found in line in the same general direction as the surveyed line. One marked tree is not proof of a survey, but can be given weight if later found to be in line. Large old trees and low value timber species like beech, hickory, and black gum are more likely to be present and have marks. Marks are usually found about breast height. The pattern and location should be consistent.

Age of Old Marks. The age of a mark can be determined within a year or less. Remove a block of wood including the mark, from the tree, and count the rings since the mark was made, each ring being one year. Rings in oak are easier to count because the spring growth consists of large pores, while summer pores are small. Pores are the same size in species like maple and birch.

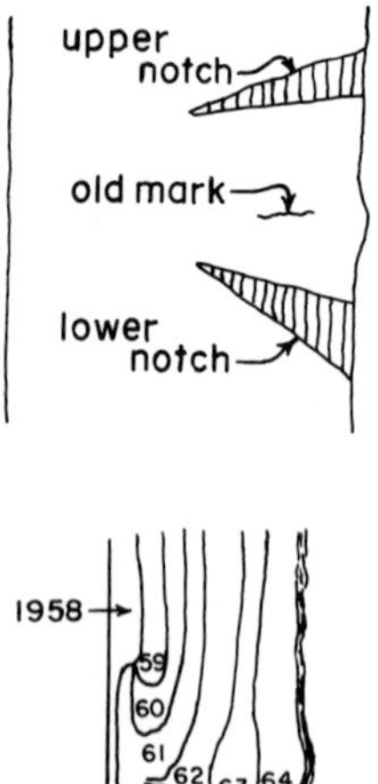

Removing the Block. Make a notch with axe or saw, above and below the mark. As the notch is deepened, watch for signs of discoloration or disturbance of growth rings. See diagram. When this point is found, continue to deepen the notches to get beyond the disturbance. Bring notches close together at the back. Then cut into the tree on each side of the mark. Continue cutting until the block breaks loose.

Counting the Rings. If a sharp axe was used, it may be possible to count the rings at once. See diagram. The cambium or growing layer will be found just inside the bark. The cambium adds a layer of bark and a layer of wood to the tree each year. Count the rings from the outside toward the point where the disturbed area stops. When the cambium is severed by an axe, the tree begins to repair the damage by putting on extra growth. By examining marks with a 2 or 3 power magnifier, the date the marks were made might be told within a few months. The block should be marked for identifying later and the notes reflect what was done. The block should be taken home.

If rings are hard to count, saw off one face and smooth with file. Brush off dust, put block under a faucet and flush out pores. Rings may be easier to count wet or when held at a certain angle under an electric light or flashlight beam. If mark is not visible, saw off another slice. Another way is to cut block in two vertically.

If there is a choice, it is better to block out a line or pointer tree than a corner tree, unless the age of the corner mark is critical, such as for testimony in court.

Safety in Surveying (See 1970 Occupational Safety and Health Act.)

Rural land surveying usually takes place in rough woods terrain. There is a good chance for accidents or injury to occur—tripping, falling, slipping, cutting a leg, getting something in the eye.

Clothes. All workers should wear proper field clothes. Shoes are most important. Rubber "lug" soles are good for firm surface contact. Six inch shoes or higher boots with lugs can be obtained at certain shoe stores. Some shoe repair shops can put on lug soles and heels. A hard cap will protect the head; light color for better visibility. A "bump" cap is good for most woods work; it is light weight, gives good head ventilation, and can be adapted for cold weather. Shirt, jacket, pants, and socks should be selected according to the weather expected. It is better to take too much than too little; an excess can be stuck through the rear of the belt. Gloves may add to comfort but usually get in the way, and in some jobs may be unsafe. Extra dry clothing in the truck will come in handy. Naugahyde-front trousers are fine protection against briers and thorns.

Warning. Before starting work the surveyor should talk with the crew about the dangers of the work and how to prevent accidents. Keep tools sharp and alert axeman to not cut themselves or into rocks. If a fall is certain, fall as relaxed as possible, dropping tools out of the way, and ease onto the ground.

First Aid Kit. A small kit should be carried. A snake bite kit is needed in areas with poisonous snakes. It is good if the surveyor has taken courses in first aid.

Repair of Tapes at Home

The author has used ordinary 2-component epoxy glue with Eureka sleeves and splices are holding. The glue is suitable for home use only because of the setting time. Burn solder and flux out of the sleeve or use an old one and clean it well. Hammer the curled ends of the tape flat and file sides a little if they widen; do not hammer too heavily. Clean tape-ends well with file or emery cloth. Make a wooden jig out of a 6″ piece of 1 x 2 stake. Saw a V-notch in one side about 2″ wide and 1″ deep. Place one end of tape on top of one side end of jig and clamp it with a 2″ C-clamp. Fill sleeve with glue and slip it over the clamped tape end open side up. Clamp the other end of tape to the other side end of jig and work it into sleeve until tape ends meet in middle of sleeve. Turn both clamps tight with tape straight and ends together; the tape should now be immovable. Add glue as needed. With pliers press sleeve snugly shut. Clean off excess glue, add a C-clamp and let it set till morning, then file off sharp projections.

Field Repair

The Eureka sleeve if fresh and properly used will make a good repair. However, a match or heat is required to melt the solder and flux, which may dry out in storage.

Inexpensive epoxy glue is now available that sets in 5 to 10 minutes and does not require heat. The author suggests that the two components be placed in separate screwtop glass vials of one-dram size or less, and carried in a small plastic box, with a few clean sleeves, a small C-clamp or spring-type paper clip and emery cloth. The splice is made about as stated above and clamped for about 10 minutes.

Fiber Glass Tape

An inquiry to the Lietz Company reveals no method of repairing their fiber glass tapes. Their 200-foot utility tape measures about 200.075 feet if tension equal to that required for a steel tape is applied. The Company advises that less tension should be applied to their fiber glass tapes. The extra length with more tension could help offset sag and error due to alignment.

Chapter 37

REFERENCE MATERIAL

Items marked with an asterisk (*) should be owned by every surveyor.

Public Works

**Manual of Instructions for the Survey of Public Lands of the United States.* 1973. Bureau of Land Management, U. S. Department of the Interior. For sale by the Superintendent of Documents, U. S. Government Printing Office, Washington, D. C. 20402.

**The Ephemeris.* For current year. Bureau of Land Management, U. S. Department of the Interior. From GPO. Order early. 80¢ in 1974.

**United States Magnetic Tables for 1960* (or latest issue). Publication 40-2. 1962. National Geodetic Survey, U. S. Department of Commerce. From GPO. May be out of print.

**Magnetic Surveys.* 1956. Serial No. 718. National Geodetic Survey, U. S. Department of Commerce. From GPO. Probably out of print.

**Isogonic Chart of the United States.* No. 3077. 1970 or latest issue. From Director, National Geodetic Survey, Rockville, Maryland 20852.

**Standard Field Tables.* 1956. Bureau of Land Management, U. S. Department of the Interior. From GPO.

**Elements of Surveying.* 1964. Technical Manual No. 5-232. Department of the Army. From GPO.

**Special Surveys.* 1953. Technical Manual No. 5-235. Department of the Army. From GPO.

Topographic Surveying. 1953. Technical Manual No. 5-234. Department of the Army. From GPO.

United States Government Quadrangle Maps. Index map free from Distribution Section, U. S. Geological Survey, 1200 South Eads Street, Arlington, Virginia 22202. Also from certain local stores.

Manual of Practice for Land Surveying. 1960 or later issue. North Caroline State Board of Registration for Professional Engineers and Land Surveyors, 1307 Glenwood Avenue, Suite 152, Raleigh, N. C. 27605.

Private Works

**Boundary Control and Legal Principles.* 1969. Curtis M. Brown, H. Frederick Landgraf and Francois D. Uzes. John Wiley & Sons, Inc., 440 Park Avenue South, New York, N. Y. 10016.

**Evidence and Procedures for Boundary Location.* 1962. C. M. Brown and W. H. Eldridge. John Wiley & Sons, Inc.

Law of Surveying and Boundaries. 1959. Clark and Grimes. The Bobbs-Merrill Company, Inc., 4300 West 62nd Street, Indianapolis, Ind. 46268.

The Legal Elements of Boundaries and Adjacent Properties. 1930. R. H. Skelton. The Bobbs-Merrill Company, Inc.

Surveying Theory and Practice. 1947. John C. Tracy. John Wiley & Sons, Inc. (Or Davis, Foote & Kelly; Breed & Hosmer, et al.)

Forestry Handbook. 1956. The Ronald Press Company, New York, N. Y.

**Bibliography of Property Survey Literature.* 1963. Winfield H. Eldridge. American Congress on Surveying and Mapping, Box 470, Benjamin Franklin Station, Washington, D. C. 20044.

Technical Standards for Property Surveys. 1946. ACSM. Free.

**Metes and Bounds Descriptions.* 1972. Jesse E. Fant, Alver R. Freeman and Carlisle Madson. From Minnesota Land Surveyors Association, Box 15001 Commerce Station, Minneapolis, Minn. 55415.

**Land Survey Descriptions.* 1970. William C. Wattles. From Title Insurance and Trust Company, 433 South Spring Street, Los Angeles, Calif. 90054.

A Surveyor's Table of Natural Cosines, Sines and Tangents. 1959. Calders, 1828 Hilyard, Eugene, Or. 97401.

**4300 Review Questions for Surveyors.* 1971. Russell C. Brinker, Sun City, Ariz. 85351. Or Technical Book Company, 253 South Spring Street, Los Angeles, Calif. 90012.

Each surveyor does not have to own all these items. It should be possible to exchange books with other surveyors. Books might be bought by a group of surveyors and placed in a public library or other place of public access. Public agencies should have books available for its surveyors.

Chapter 38

CASE HISTORIES AND PROBLEMS

Professional workers are often reluctant to discuss the procedures used to solve a problem or perform a job. A doctor does not often tell a patient how he diagnosed an ailment; nor does an attorney explain to a client the complex laws involved in his opinion. This reluctance is often soundly based. However, surveyors have shared in this reluctance to a degree inconsistent with their responsibility to the client and public. A surveyor should, for example, in most cases prepare a report telling how he reached his conclusions.

Land surveyors have no final authority to fix corner or boundary line locations; they are fact-finders, interpreters, recommenders, or arbitrators. The solution of a case is subject to adjustment if new evidence is found, if parties disagree, or if the Court so decides. A surveyor's reputation, like a judge's, may often be gauged by the number of times his judgment is overturned.

This chapter is devoted to actual examples of survey problems, mostly in West Virginia, stripped to their bare essentials, and how they were solved. The solution in other states can be similar, but the surveyor there must decide. A rule or court decision covering the situation will not always be found, and this is where the surveyor's judgment is required. Some solutions are not 100% perfect, nor will every surveyor agree with them.

Examples are divided into two groups: *Rural* and *City*. In addition to the abbreviations listed in the Preface, these are used in this chapter:

BPO—being part of	N/C—nail and cap
STB—said to be	H/T—hub and tack
HD—horizontal distance	Bl.—block
SD—slope distance	m or l—more or less

When distances are shown only by figures (such as 124), they mean poles (rods); other units of length are stated, usually in feet ('). The degree and minute symbols are omitted. North in diagrams is toward the top of the page unless otherwise shown. Diagrams are not to exact scale.

Rural

—A—

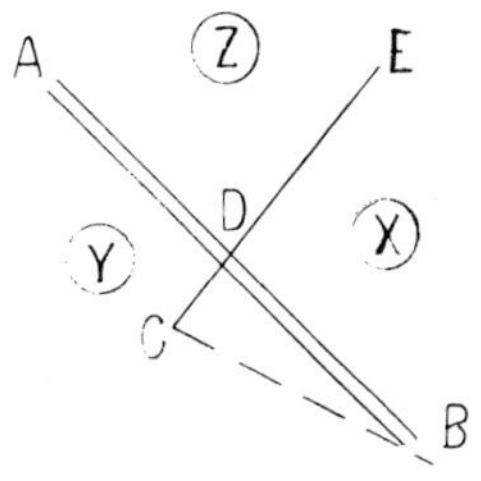

GIVEN: Road AB. Corner marker at E in deeds X and Z. Deed Z calls for stake in road (D). Deed Y calls for line BC to follow road. Deed X calls for road as line.

FOUND: Call EC in deed X carries 4 poles SW of road. Owner X claims he owns land SW of road because his poles reach to C. No marker found at C.

SOLUTION: Owner X goes only to road because monument such as a road takes precedence over course and distance.

—B—

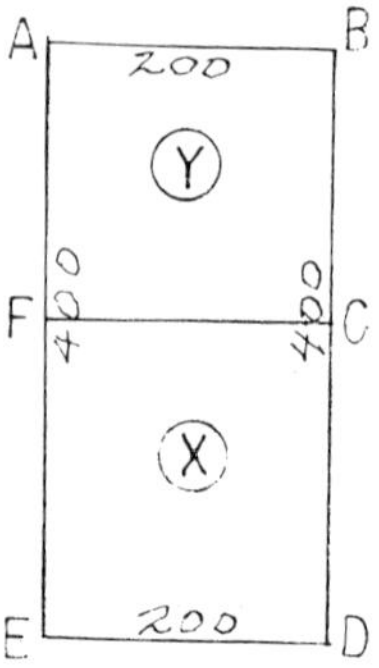

GIVEN: Deed X says: "Being south half of a 500 acre tract", with distances on plat being surface measure.

FOUND: Authentic markers at B, D, and E. BD measures 441.4 HD and DE 207. Evidence of old fence near F, 205 from C. Nothing at A.

PROBLEM: Establish location of C and F.

SOLUTION: Actual distance BD takes priority over deed call. Establish C halfway between B and D; establish F by intersecting 205 from C and 220.7 from E.

—C—

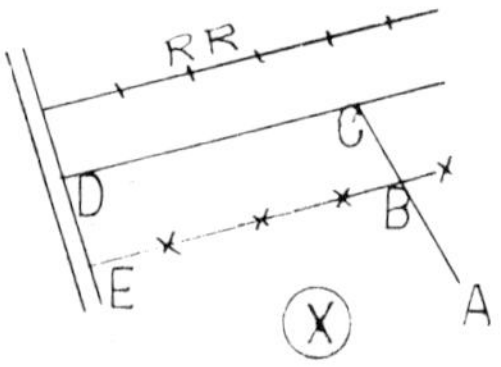

GIVEN: Junior deed X calls: "Beg. at a hickory (A) then N 32 W 71′ to post in south edge of railroad r/w (B or C), and with same 247′ to stake on east side of —— road in south line of railroad r/w (D or E)." Senior deed for railroad calls: ". . . N 74-40 E with center line of a strip 50′ wide on each side of center line for 600′, then 100′ wide on each side. . . ."

PROBLEM: Locate line of deed X along RR r/w.

FOUND: Railroad was a tangent in a tunnel, the mouth of which is 100′ NE of X's land; hickory at A; old intermittent fence along BE: AB is N 31-55 W 72′; BE measures S 76-20 W 249′; road is abandoned.

SOLUTION: Set up in center of RR at mouth of tunnel and transfer RR center line to BE area; found that track center line is about 93′ from BE. RR Co. engineer states that "a 6-strand wire fence was located 93′ from center of tract in 1917 . . . I feel that our r/w line is 43′ north of this old fence. . . ." Conclude that r/w is along CD and that fence BE was for convenience only. Line AB must therefore be extended to C, making AC 118′. Survey CD as S 76-15 W 247′ parallel to railroad, and select best location for D that checks with road edge.

—D—

GIVEN: Deed X says: ". . . westerly to a point in the middle of —— Creek, thence down said creek to its junction with the river, thence leaving said Creek easterly. . . ."

FOUND: Recent resurvey of said line for conveyancing is described as: ". . . N 82 W to a point on the east bank of –– Creek, thence along the east bank N 14-30 E 290′ to a stake on the east bank, thence N 9 E 120′ to stake on the east bank, thence leaving said creek N 89 E. . . ."

PROBLEM: The recent description leaves a strip of land unconveyed between the center of the Creek and the east bank, although the grantor says he intended to convey it.

SOLUTION: Revise description to read: ". . . N 82 W to the center of –– Creek, thence down said creek N 14-30 E 290 to a point, thence N 9 E 118′ to a point in said Creek, thence leaving said creek. . . ." Note: Make good witness markers on bank.

–E–

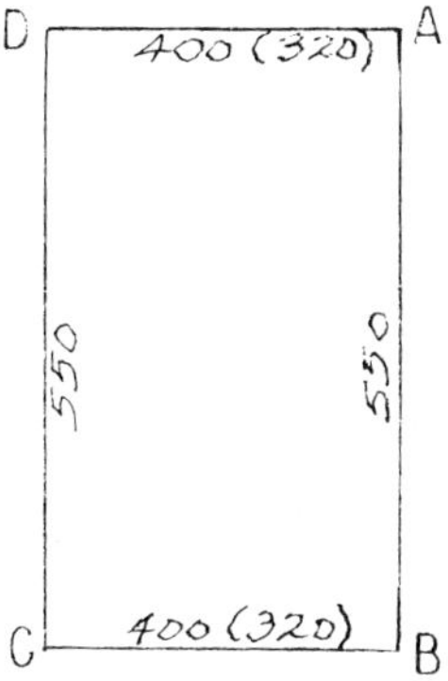

GIVEN: Delinquent land tax sale deed for 143 acres "being the residue of a 1375 acre tract sold by Y to Z in 1887." Calls of 1375 acres are 400 by 550.

PROBLEM: Where is the land.

FOUND: All tracts sold by Z since 1887 are in area ABCD (320 x 550). All parcels along the exterior of ABCD called for tracts to adjoin the "1375" acres. ABCD is similar in shape to other adjoining tracts of a large survey.

CONCLUSION: The "1375" acre tract contains only 1100 acres. Probably had to yield to adjoiner encroachment. 143 acres does not exist.

–F–

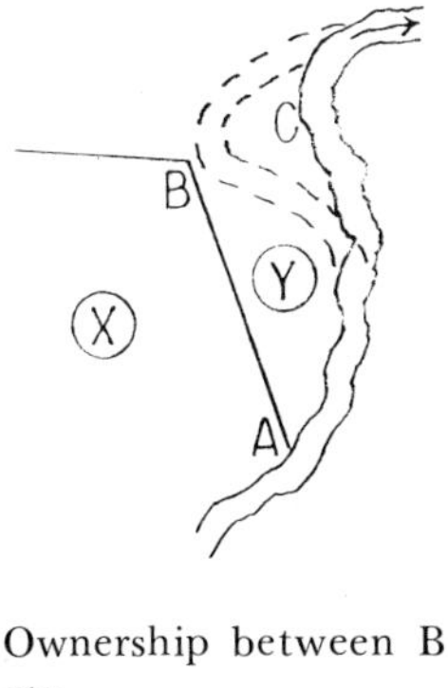

GIVEN: Old deed X calls for A to be "two willows on bank of river"; B is "locust on the river bank."

FOUND: A is on river bank but B when platted is 10 poles from present bank. Traverse closes fairly well.

SOLUTION: Aerial photo shows that river had an old channel touching B (like dash lines); confirmed by ground inspection. Ownership between B and C depends on rapidity of channel change, etc.

–G–

GIVEN: Deed X calls: "Beg. at white oak (D), thence N 83 W 44 to double beech in flat bottom (B or C), N 45¼ E along Z's line 123 to E. . . ." Deed Y calls; ". . . Beg. at A, then S 83 E 75 to stake pin oak pointer" (D). Deed Z calls: "Beg. at A, S 83 E 36 to double beech in flat, then N 45¼ E 123 to E."

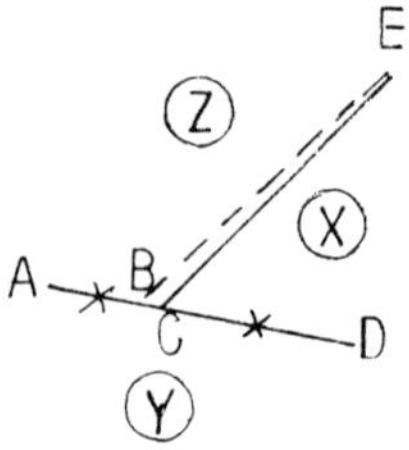

FOUND: Authentic markers at A, D, and E. Fence along line ACD and along part of CE. Fence intersection in flat at C with decayed beech log nearby. DC measures N 80 W 39.01; CE is N 47 E 125.8; AC is 36.2. Variation of 3° found on line DC, if applied to CE, would make it N 48-15 E, which would support a corner nearer B and check closer to 44 call in deed X.

PROBLEM: What is best location for beech corner.

SOLUTION: Majority of evidence supports location at C. Variation and one distance not enough to predominate.

–H–

GIVEN: Deed X calls: ". . . S 46½ E to white oak on north bank of road, then N 70½ E 52 to span oak on north side of road, S 6 E 28 to two locusts on south side of said road. . . ." Adjoiner deed Y has same calls.

FOUND: Calls generally follow road; no white oak or spanish oak found. Locust stump found.

PROBLEM: Resurvey the lines and update the description.

SOLUTION: Obtain owners' agreement for line to be the center of road. Extend S 46½ line to center of road; run new lines up center of road. Write new description "with meanders of road", and state "as agreed between —— and ——."

–I–

GIVEN: Owner X conveys to Z 166½ acres "part of 177½ acres, 11 which were conveyed to Y." Deed X has calls for 177½ acres.

FOUND: Y's successor W claims about 9 acres of the 177½ acres projecting over a road where W has other land. Part of 9 acres was occupied and used by Y for longer than adverse possession period; the 9 acres is the only land of X that adjoined Y. No recorded deed found from X to Y.

PROBLEM: Where is the 11 acres and who has better title.

SOLUTION: Weight of evidence shows that 11 acres would have to be where the 9 acres is located. Resurvey the calls of the 177½ acre deed in vicinity of 8 acres; make lap of the 9 acres. Obtain statements of possession; turn data over to attorney for attention.

–J–

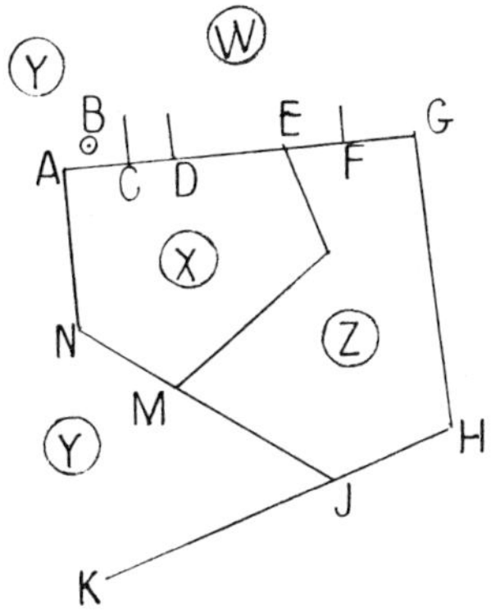

GIVEN: 1907 senior deed X says: "Beg. at poplar (A), then N 83 E 148 to stake (E), S 23 E 89 to stones, S 47 W 117 to stake on originaline, N 61 W 90 to stones, N 8-45 W 102.92 to beg." Deed Y (1927) says: "from (C) S 85 W 45 to a beech, S 8-45 E 102.92 to chestnut, S 61 E 221 to sugar (J) in line HK and with it S 70 W. . . ."

FOUND: Stone (G) STB corner to parent tract in deeds W and Z from which 1907 deed came. Stones as called for at D, E, and F. At A old stone set beside 26″ poplar with one blaze facing stone. B is unmarked beech 10 easterly of A and 4 north of AC and N 84-50 W of C in 1964. N is small stone pile 99.0 HD from A and 100.7 SD from A. AG in deeds total 281. C is at west edge of RR r/w. Excessive errors of closure in all deeds.

PROBLEM 1: Is A or B better corner location.

SOLUTION: Run random line from G on S 85 W, which passes within a few feet of F, E, and D. Extend GFED past C and intersect NA extended 1.3 northerly of A; then correct line GFED passes 0.4 south of C, 3.6 south of B and totals 279.9 HD and 288.8 SD. Stone and poplar is superior to beech.

PROBLEM 2: Why not use a point 1.3 northerly of A as corner.

SOLUTION: Called for monument (poplar controls over course and distance. Location A leaves less grounds for owner Y to complain.

PROBLEM 3: CB measures N 89-50 W in 1964. What should 1927 S 85 W call of adjoiner be in 1964.

SOLUTION: Variation is 0-35; bearing should be S 85-35 W.

—K—

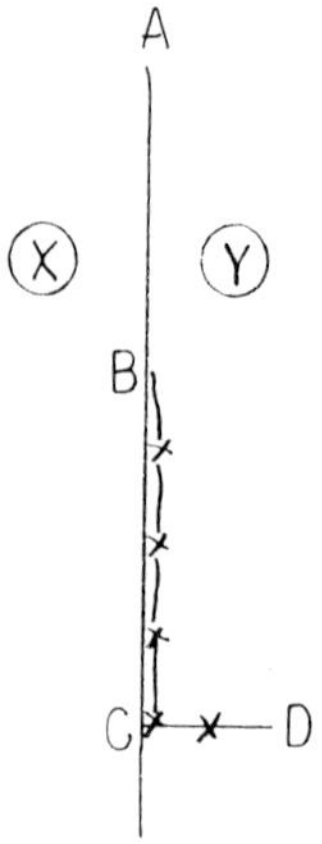

GIVEN: Line ABC 2400′ long. Agreed corner at A. Good old fence on line AB considered to be line fence; new fence along BC which client Y says is not on line. White oaks called for at C and D.

FOUND: Unmarked large white oaks at C and D with fence attached to each; distance CD checks within 3″ of deed call distance.

PROBLEM: Adjoiner X built new fence east of BC, ending it 8′ east of C. Where is correct line.

SOLUTION: Run AB and extend it; found it hits C. Advise owner X that his fence is on wrong location and explain why.

—L—

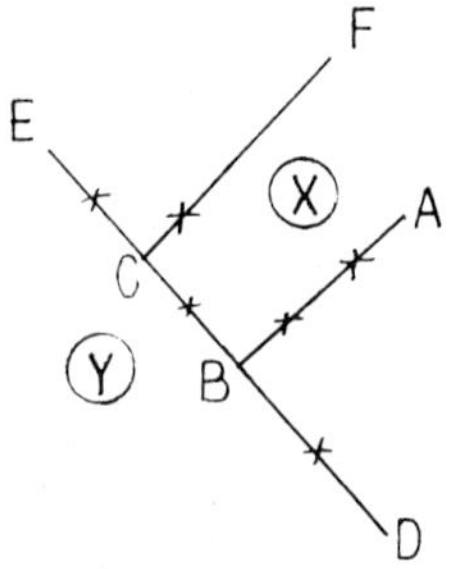

GIVEN: Deed X calls: "Beg. at beech (A) then S 48½ W 18½ to stake and pointers in Y's line (B), and with his line N 49 W 64 to hickory now down (C) then N 40 E . . . to (F). . . ." Deed Y calls: "Beg. at sugar (D) then N 50 W 137 poplar (E)."

FOUND: Beech at A. No stake, hickory, poplar, or sugar; good old fence fairly straight along DE; remains of old fence between A and B; its bearing S 49-19 W; distance from A to fence DE is 18.76.

PROBLEM: Locate corners B and C.

SOLUTION: Obtain concurrence of owner Y to consider fence BE to be about on line. Set marker at B which is S 49-19 W 18.76 from A; run from B, N 45-04 W with fence; at 63.6 find junction of old fence toward F; set marker there (C). Variation in line AB (0-49) and BC (3-56 and 4-56) indicate surveys made at different times.

—M—

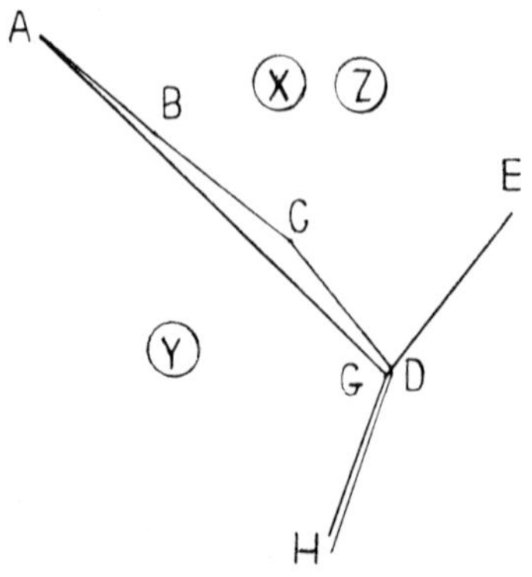

GIVEN: Deed X calls: "Beg. at 2 chestnuts on ridge (A), then S 50-45 E 904′ (to G), N 49½ E 972′ (58.9 poles) to (E). . . ." Deed Y covers "the land south of — ridge, the land north of ridge having been sold to Z." Z is predecessor to X. Successor to Y had land surveyed on ridge ABCDH.

PROBLEM: Which is correct—AG or ABCD.

FOUND: Deed Y is senior to deed

X. Found 2 chestnut stumps at A; fence along DE about 53 poles; no fence south of D.

SOLUTION: Monument on ground (such as ridge) in senior deed takes priority over course in deed X. ABCD is correct.

—N—

GIVEN: Deed calls: "from A to beech on bank (B) then down the creek."

FOUND: Both A and B; fence runs down creek bank from B. Adjoiner deed calls for the creek.

PROBLEM: Where is client's true corner at B and how to locate it.

SOLUTION: Extend AB to center of creek; survey down creek center.

—O—

GIVEN: Deed X calls: "Beg. at birch and lynn on —— Creek, then N 15 E 7 to stake in the creek, then down said creek with its meanders N 15 W 14, N 63 W 55 . . . to falls in creek. . . ." Older deeds in chain give second calls as N 53 W 14. Adjoining deed calls: "Up —— Creek. . . ."

FOUND: Agreement (recorded) between owners X and Y calls for set stone at or near birch and lynn. Found set stone with 12″ lynn pointer with old blaze; no old birch or old lynn. 12′ high falls in creek.

PROBLEM: What is proper line between parties.

SOLUTION: Run deed X calls up creek from falls and tie to set stone; found 4½° variation; from stone to creek center being N 19½ E 7, and first call down creek being N 48½ W. Second call in deed X (N 15 W) was typographical error because N 48½ W plus 4½ is N 53 W as in older deeds. Chiselled a mark on rock in creek at end of N 19½ E 7 call; actual creek meander is line between parties.

—P—

GIVEN: 1846 call "S 80 E 126 to red oak on bank of —— run." Adjoiner deed has same call.

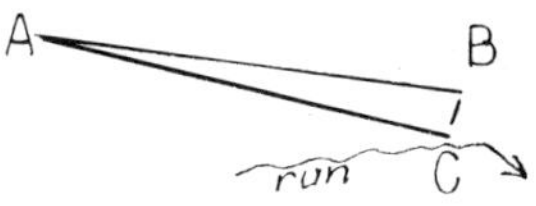

FOUND: Monument at A undisputed by 4 adjoiners. Variation from 1846 to 1965 is 5-25, making current bearing S 74-35 E; AC bearing in 1965 is S 71-45 E; no oak found at B or C. No other markers close enough to help. BC measures 6½ poles.

PROBLEM: Should correct line be AB or AC.

SOLUTION: The bank of this Run could hardly be 6½ poles from its center. As monuments called for and found take priority over course and distance, AC would be the better location; select a point on bank north, 126 HD from A, about 10′ from center of creek.

–Q–

GIVEN: New parcel to be surveyed on east side of –– Creek.

PROBLEM: Corners to be established to make the bank the line and not the center of the creek.

SOLUTION: Set marker on east bank above high water mark; set the other marker the same way. In description, call for "a –– on east bank of –– creek, about ––′ from center thereof, thence with said bank S 28-35 E 184.7′ to a –– on east bank of said creek about 45′ from center therof, . . ."

–R–

GIVEN: Deed X calls: "Beg. at stake in center of the road (A), then S 14-30 W 16 (to B) S 67 E 10, N 14-30 E 16, N 67 W 10 to beg. . . . BPO deed Y." Deed Y calls: "Beg. at maple on bank of –– Creek (E), then S 70 E 110 to white oak on a ridge (F). . . ." Deed Z calls: ". . . to large oak (F), then N 70 W 22.32 to white oak on east limit of said of said Pike (G), then with said Pike N 21½. . . ."

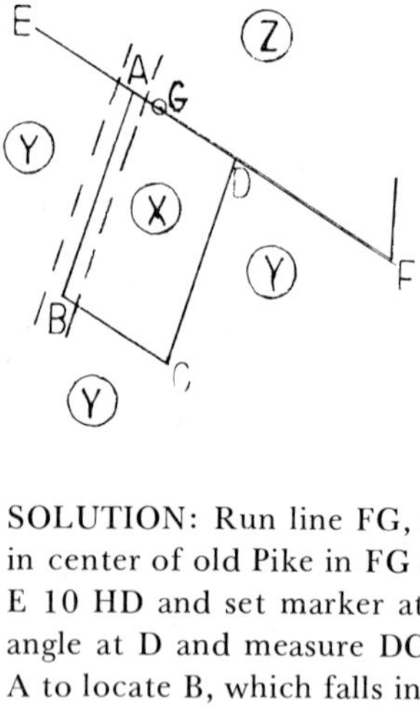

PROBLEM: Locate corners of tract X.

FOUND: Old White oak stump at F; 24″ white oak stump at G. Old fence along FG: sign of old turnpike along AB. AG measures about 13′.

SOLUTION: Run line FG, found it to be N 66-17 W 22.55. Establish A in center of old Pike in FG extended 0.8 pole. Measure from A, S 66-17 E 10 HD and set marker at D. Angle ADC in deed X is 98½; turn this angle at D and measure DC as 16 HD: intersect 10 from C and 16 from A to locate B, which falls in or near center of old road.

–S–

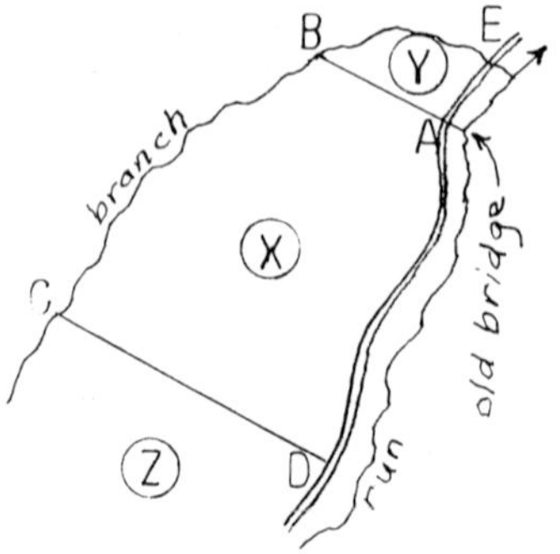

GIVEN: Deed X calls: "Beg. at point on west side of State Road, N 63-10 W –– feet from lower abutment of the old concrete bridge across –– Run, then N 63-10 W parallel with old –– line about 60 to a branch (B), then up said branch with its meanders about 176 to corner of Z's land (C) then S 66-30 E about 117 to said State road, then with said road about 170 to beg." Deed Y's calls for line AB is same as in Deed X.

PROBLEM: Where is line between X and Y.

FOUND: Fragments of old bridge abutment; markers at C and D. DA measures 151. Nothing at B. Owner X measured 170 from D which went to branch at E, so he contends his line should be branch EBC. EB measures about 60; BC measures about 157. Deeds X and Y written without field survey.

SOLUTION: Poles must give way to monuments found. Obtain owners' agreement on specific starting point at lower abutment. Find present bearing of old "N 63-10 W —— line" in vicinity and run AB on its bearings.

—BB—

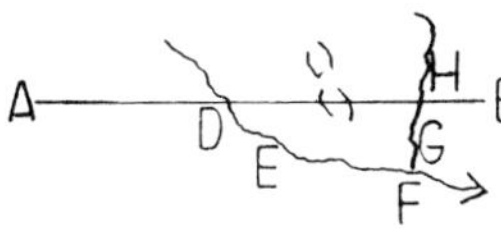

GIVEN: Line AB to be surveyed and marked goes over very steep ridge with cliffs; very difficult job.

FOUND: Nice flat for surveying along nearby stream.

SOLUTION: Run line AH. Run traverse DEFGH, making GH near to right angle with AB; compute closing call HD. Compare closing call HD with bearing of line ADH. Figure difference as angle A of right triangle (p. 218 SFT). Calculate side "a" with formula "a=c sin A. Adjust length of GH to get in line AD extended. Finish random line to B, figure offsets and set stakes on true line. Traverse DEFGH furnishes check on the difficult part of line AB. Alternate Solution: Large scale plat would avoid calculations.

—T—

GIVEN: Deed X calls: ". . . S 70 E 198 to stake (A) in line BC on north side of road, then N 33½ E. . . ." Deed Y calls: ". . . meandering the road northwesterly to stake in line BC, then N 30 E with line AC. . . ." Deed Z calls: ". . . S 66¼ E 116 to 2 white oaks, then S 32 W 50 to beg., a white oak near —— Run. (B)."

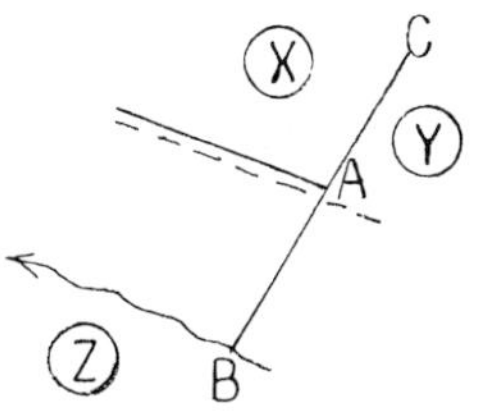

FOUND: Old road and authentic white oak (b) at Run; 2 white oaks not found. AB measures 40. Deed Z does not close, last line being too long.

PROBLEM: Owner Z contends his corner A should be 10 poles toward C to get his 50 poles distance.

SOLUTION: 40 is correct length of Z's call because monument found takes priority over call; closing call is likely to contain error.

—U—

GIVEN: Deed X calls: "... to an iron pipe (A), then S 20-45 W 141′ to an iron pipe (B) 14′ north of center of a 20′ alley, then N 79-30 W 150′ to iron pipe (C) 14′ north of center of a 20′ alley, then N 68-05 W 864.3′ to iron pipe (d) 14′ north of center of said alley."

FOUND: Pipes at A, B, D, apparently undisturbed; angles and distances OK. Pipes on south side of alley not in good alignment.

PROBLEM: Pipes 14′ north of center of 20′ alley leave questionable ownership of 4′ strip. Who owns this strip.

SOLUTION: Turn angles and measure distances in Deed X and set pipe C. Furnish data to client to obtain attorney's opinion on the title.

—V—

GIVEN: Deed Z calls for 2 sugars at a bear wallow (D) and a straight line to E. Deeds X and Y call for a poplar at C and come from a common deed source.

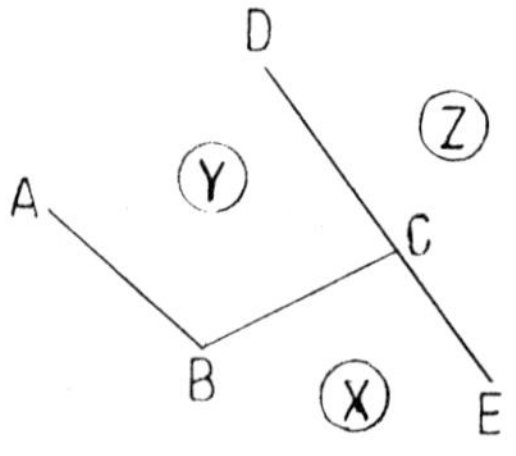

FOUND: Corners at A, B, E. One sugar at D; two old poplar stumps near C—one about 50′ SW of the other. D is visible from A and B but not from C or E.

PROBLEM: Which poplar stump should be used.

SOLUTION: Run lines ABCE. At A and B and from points between A and B take shots on D. Establish D on plat by intersections and draw line from D to E. Select that stump which is closest to line DE regardless of distance from B to C.

—W—

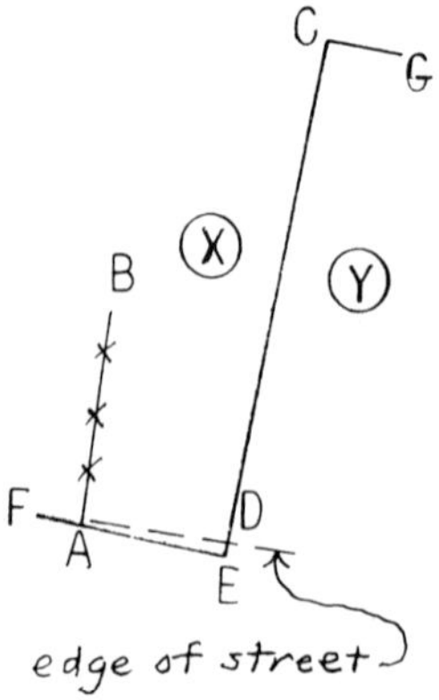

GIVEN: Deed X calls: "Beg. at stake (A) on NE side of —— Street, then N 7-45 E 122′ to post (B) ... to an iron pipe (C), then S 11-50 W 312′ to IP at —— Street (D) then with edge of street 80′ to beg."

PROBLEM: When platted, last call goes 8′ beyond A making large error of closure (1 in 213). CD measures 306½′ (5½′ short). What to do about it.

FOUND: Fence corners at A and B and good old fence along AB which is 121½′. Found pipe at C, pin at D and tree called for at G. Pipe at C believed set after original survey. A and D about 15′ from center of unpaved street (once a country road legally 30′ wide). AD measures 80′. Deed Y not helpful.

SOLUTION: Obtain agreement of X and Y owners to set new C marker 312′ from D and deed distance from G.

–X–

GIVEN: Deed X calls: "Bg. at 3 lynns (A), then S 67 W 135 to lynn and sugar (B), S 68½ E 78 to beech (C). . . ." Deed Y calls: "Beg. at 3 lynns 6 poles south of a run (A), then S 67 W 280 to maple and birch on a ridge (D). . . ." Deed Z calls: "Beg. at birch and maple on ridge (D), then N 67 E 144 to lynn and sugar (B), S 68½ E 75 or 78 to beech (C). . . ."

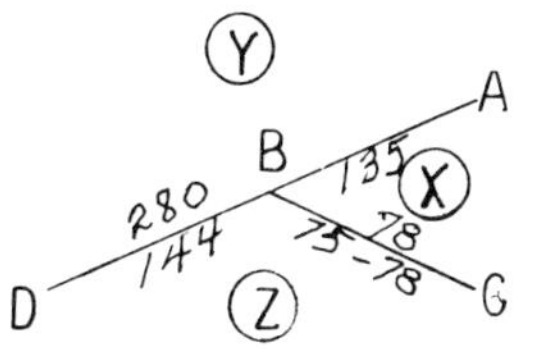

FOUND: At A, intersection of old fences about 6 poles south of run. At B nothing. At C nothing. At D down birch and maple on ridge.

PROBLEM: Locate corner B.

SOLUTION: Run line AD; it is 281. Prorate distance as follows: 279:135; 281:AB, or 279 AB equals 37935 and AB equals 135.97. Establish B in line AD 136 from A. BD then is 145. Alternate solution: give senior deed its distance; junior deed takes remainder.

–Y–

GIVEN: Line AB to be surveyed. A and B are identifiable.

PROBLEM: Mark line so it can be run many years later without a new random line missing the corner very far.

SOLUTION: Run random line A toward B. Calculate error of closure and offset stakes to true line; set a pipe at first station from A toward B and describe it in survey. Future surveyor can then set on A, sight on pipe, determine its bearing and extend that line.

–Z–

GIVEN: 1878 deed calls: "Beg. at gate post by the old road in —— line, then with an agreed line S 4 E 24 to stone pile where chestnut sprouts called for, an agreed corner. . . ."

FOUND: Gate post on north side of road STB corner; old stone pile S 2-42 E 20.92 from gate post. Variation from 1878 to 1965 should be about 3½ instead of 1-18. Difference in distance and variation indicates stone pile may not be at site of original chestnut sprouts. Adjoiner deeds shed no light.

PROBLEM: Should corner be at stone pile or should a new corner be made S 2-42 E 3.08 from stone pile, or where.

SOLUTION: Client and adjoiner have no objections to either the post or the stone pile. As 1878 deed indicated stone pile was agreed many years ago, it should be accepted now.

—AA—

GIVEN: Deed X calls: "Beg. at small ash, then SE 59 to maple and white oak pointer on north bank of road, then up road (various calls total 177.2 poles) to small white oak, then N 21 E 40 to stone. . . ."

FOUND: Clump of ash N 44-50 W 59 from white oak stump on north bank of road. 20″ white oak at fence corner on south side of road 182.9 from stump and S 25-40 W 40.8 from a set stone. Crooked road could account for some of discrepance in road distance. Adjoiner deeds not helpful.

PROBLEM: Are stump and white oak satisfactory corners.

SOLUTION: Distances, bearings with variations check closely enough to use them as pointers to corners in center of road.

City

—CC—

GIVEN: 1919 deed X calls: "Beg. at the hard road and First Street (A), then with Street 190′ (to B), then almost due south 165′ to stake (C), then west a straight line 165′ to pear tree at county road (D), then north with said road to beg." 1921 deed Y calls: "Beg. at hard road and First Street and with Street 190′ to stake, then almost due south 60′ to stake (E), then parallel with Street to county road (F) and with it to beg." 1925 deed Z calls: "Beg. at original corner of —— Addition (G) then westward 147′ to stake, corner of Lot Y, and with Y 60′ to stake, then parallel with first line 147′ to stake (H), then a straight line to beg."

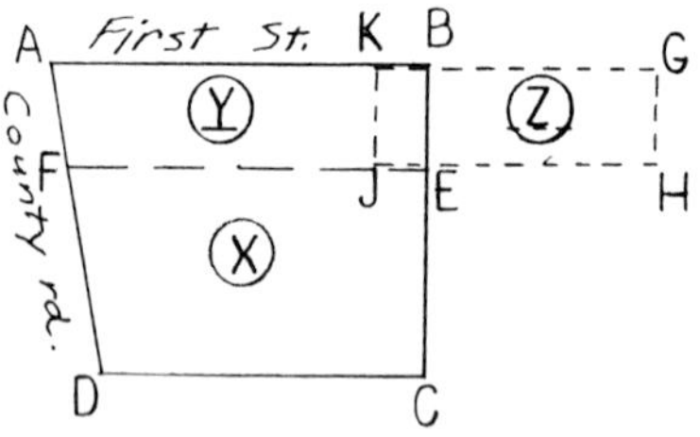

FOUND: Concrete marker at G; stump of pear tree at D. Street and road fairly well defined. Deeds X and Z from same grantor. Grantor in deed Y is grantee in deed X. Overlap of 14′ between deed Z and the other two deeds, covering BEJK.

PROBLEM: Establish line between Z and the other two owners.

SOLUTION: Deed X is senior so Z must take the lost of 14′.

—DD—

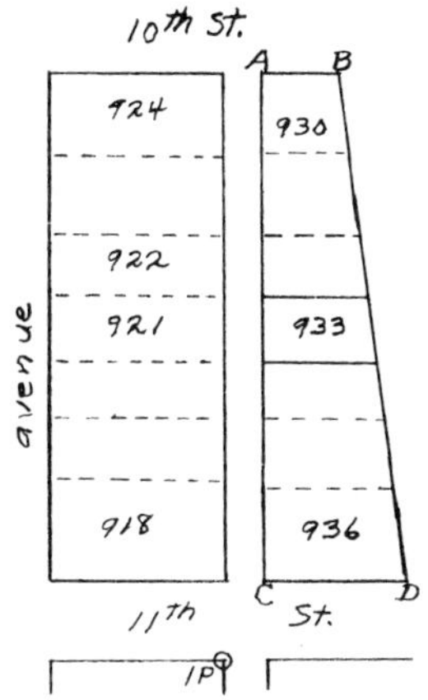

GIVEN: Plat of subdivision showing Lot 918 and 924 to be 50 x 120′; lots 919 to 921 to be 40 x 120′; lots 930 to 936 widths are 40′. AB is 37′ CD is 120′. No other lot depths shown.

PROBLEM: Locate corners of Lot 933.

FOUND: No authentic markers in area. 10th St. well curbed and location OK; 11th St. not curbed. Avenue not curbed. Found pipe near south side 11th St. Good surveyor recently set markers at corners of Lots 921 and 922; strong steel fence connects them. From 10th St. to Lot 922 is 1.2′ excess; from 11th St. to Lot 921 is 0.5′ excess. From 10th St. to pipe on south side of 11th St. is 1.7′ excess. If 1.7′ was prorated over all lots, south line of Lot 933 would be 1.8′ north of its plat location, and north line of Lot 933 would be 4.2′ north thereof and close to house on Lot 932.

SOLUTION: Use markers for Lots 921 and 922 as guide instead of inconclusive street locations. Lot 933 then is 40 x 78.5 x 41.2 x 88.3. Let streets or perhaps end lots absorb excess, if any.

—EE—

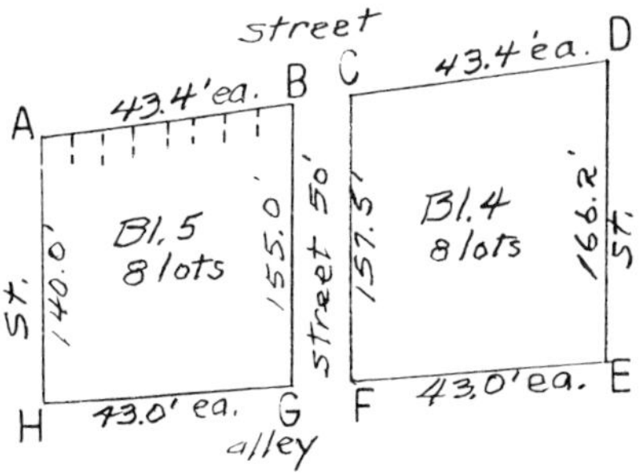

GIVEN: Recorded subdivision plat with 16 lots. AD and HE seem to be straight lines.

FOUND: Markers at ADEH reasonably acceptable. 50′ street is 51.1′ on diagonal at BC and 50.8′ at GF. Plat figures show convergence of DA toward EH is 26.2′. Distances AD, DE, EH, and AH check with plat. Then 26.2′. prorated over front of 16 lots and 51.1′ street is 0.035′ per foot (16 x 43.4 is 694.4 plus 51.1 is 745.5 divided into 26.2). CF thus should have been 154.05 (8 x 43.4 is 347.2 x 0.035 equals 12.15 from 166.2) and not 157.5. BG in same manner should be 152.26.

PROBLEM: Locate corners B, C, F, and G.

SOLUTION: All lots should share proportionately in surplus or deficit.

As only 154.05 is available for CF between streets, the original surveyor seems to have erred in calculating lot depths. Locate B and C by measuring 347.2′ and 398.3′ from A toward D; locate B and C by measuring 344.0′ and 394.8′ from H toward E. Line B, C, G, and F in with transit. Check CF for 154.05′ and BG for 152.26′.

—FF—

GIVEN: 1953 deed for Lot XY says: "S 19 E 95′, N 62½ E 50′ to stake (E) on south line of stable lot, N 19 W 95′, S 70 W 50′ to beg." Lot X described in 1903: "S 19 E 95′, N 68½ E 30′, N 19 W 95′. S 70 W 30′ to beg." Lot Y in 1910 deed is: "Part of stable lot with 20′ frontage and extending back to an even width of said stable lot." Deed Z in 1942 says: "Beg. at NE corner of Lot Y, then southerly 95′, northeasterly 60′, northerly —′, southwesterly 60′ to beg., BPO stable lot."

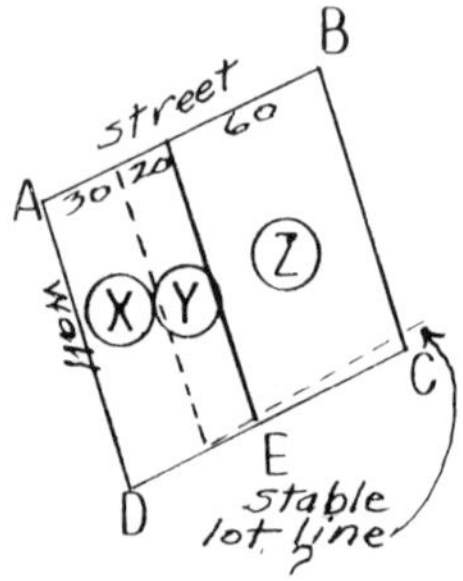

FOUND: Street OK along line AB. Concrete wall along AD recognized as line. No other markers. Weak evidence that south line of stable lot is 3.5′ north of C and 1.5′ north of E.

PROBLEM: Locate ABCD. Lots X, Y, and Z now owned by one person.

SOLUTION: Establish A and B at half-width from street center and 110′ apart. Establish D 95′ from A at wall. Decide that intent of deeds and weight of evidence is for BC to equal AD. Intersect lines 95′ south of B and 110′ east of D to locate C.

—GG—

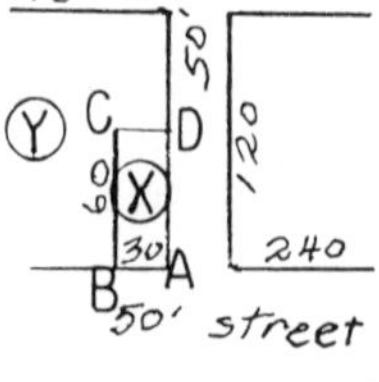

GIVEN: Lot X on subdivision plat is 30 x 60′; deed says 38 x 60′; city zoning map says 40 x 60′.

PROBLEM: Locate corners of Lot X.

FOUND: All streets paved and curbed. No authentic markers in area; houses on X and Y lots only 2′ apart.

SOLUTION: Establish A by intersection of calculated footages from center of streets. While setting pin found pipe 4″ below surface. Set pin at B, 30′ from A and 25′ from street center; B is midway between houses. 38 and 40 look wrong. With transit at A turn 90° from B and mark D, 60′ away and 25′ from street center. Mark C by intersection.

–HH–

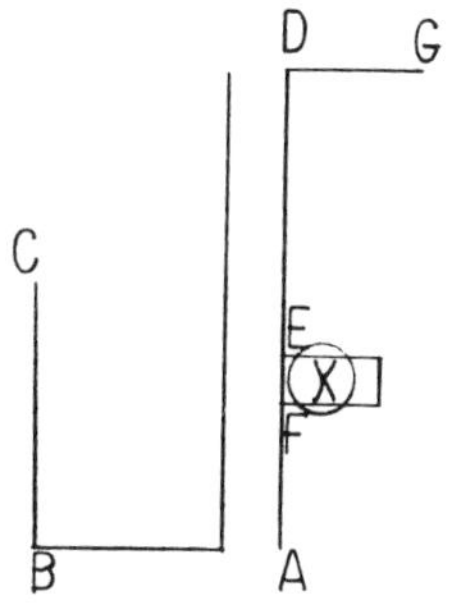

GIVEN: Subdivision plat shows markers at A, B, C; AB and BC to be 210′. No marker north of D.

PROBLEM: Locate front corners of Lot X at E and F.

FOUND: Concrete markers at A, B, C. Iron pipe at D at plat distance from pipe at G. AB is 210.06′. C to line AD is 210.1′.

SOLUTION: Measure proper plat footage from A toward D and set F and E.

–II–

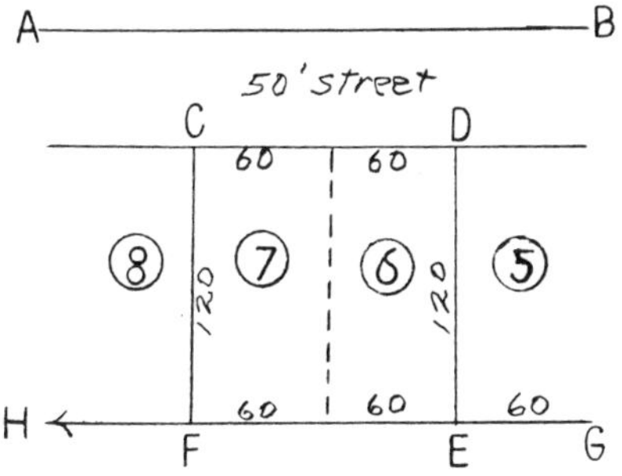

GIVEN: Lots 6 and 7 on subdivision plat.

FOUND: Undisputed concrete markers at ends of block on north side of street (AB); nothing along south edge. Concrete marker (not shown on plat) near E is 60.0′ from G and 169.55′ south of AB. G is concrete marker 170.0′ from B. IP near F is 120.22′ from E and 170.23′ from AB. IP at H (Lot 13) is 171.8′ from AB.

PROBLEM: Set markers at CDEF.

SOLUTION: Assume A, B, and G are OK, and E and F wrong. Establish E at 60′ from G and 170′ from AB. Establish F at 180′ from G and 170′ from AB. Turn 90° from E and F and measure 120′; set markers at C and D. Check CD as 120′ and to AB as 50′.

–JJ–

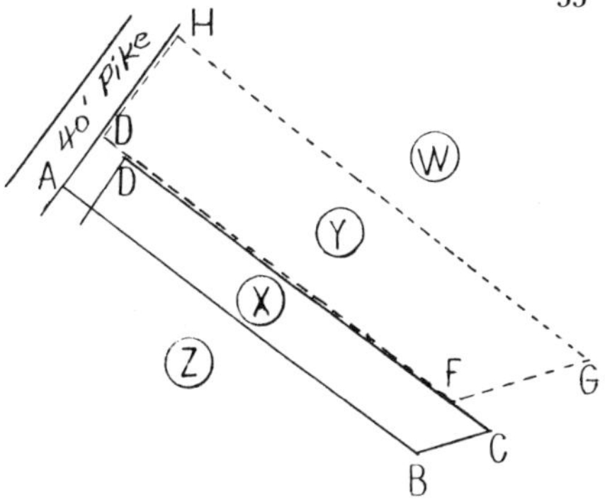

GIVEN: In small rural community, 1906 deed X calls: "Beg. at stake on east side of Pike, corner to Z (at A), and with Z's line S 54 E 22 to stake (B), N 75 E 3.75 to stake (C), N 54 W 22½ to pike (D), and with said Pike S 37 W 3 & 5/8 to beg." 1907 deed Y calls: "Beg. at corner to X on east side of Pike, S 54 E 22 to bed of old Creek, and with said bed N 75 E 6 p. & 23 links to stake, N 53 W 26 to Pike and with said Pike to beg."

PROBLEM: Owners X and Y want their common corner CF located. They can't agree.

FOUND: Calls fail to close on plat or to fit against Pike, which is 40′ wide. Same grantor in all deeds; deceased. Seniority in this order: Z, X, Y, W, V. Old hedge dividing X and Y was removed some years ago with no trace. Iron pipes at A and D 3 & 5/8 apart and 19½′ from approx. center of Pike. Fence post near B is 22.05 from A but owners non-commital on it. X and Z agree on their line within ½-1′. Old fence post at G agreed upon by V, W, and Y as corner. Pike to G measures 26.7. Nothing at C, F or H. Old creek bed along and near back line of X and Y. No witnesses who wanted to testify as to F or C location.

SOLUTION: Survey all parcels from same grantor in area. *Independent Resurvey* indicated. Measure 22 from A toward B and set marker and measure from it to post G. Prorate that distance between X and Y according to the length of their respective back lines and set marker. X would not accept this solution and sued Y; Court supported surveyor's recommendation.

—KK—

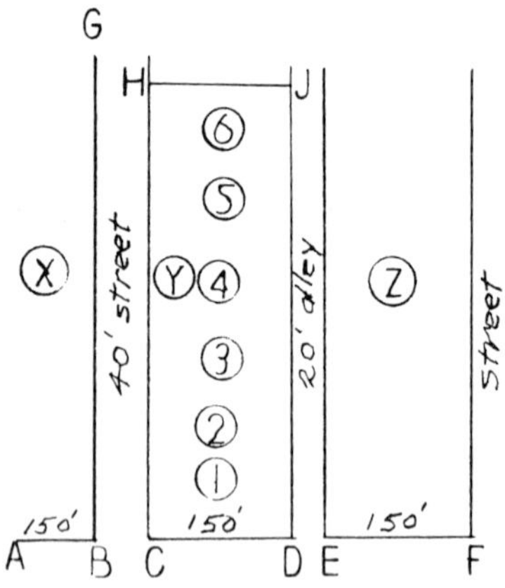

GIVEN: Lots 1 to 6 in undeveloped Block Y of Subdivision. Plat shows markers at A and F.

PROBLEM: Locate corners C, D, H, J.

FOUND: Concrete markers at A and F; 5 or 6 iron pipes at lot corners on line BG. Shortage of 6″ between B and F. Section X developed; section Z not developed.

SOLUTION: Let Lots in Y and Z absorb the shortage; make each tier 3″ short. Let new plat show 149′9″ depth for each tier.

—LL—

GIVEN: Subdivision plat shows concrete markers at A, E, F; AE is 150.4′; EF is 40.2′; angle at A is 78°. Deed X calls: "Beg. at concrete

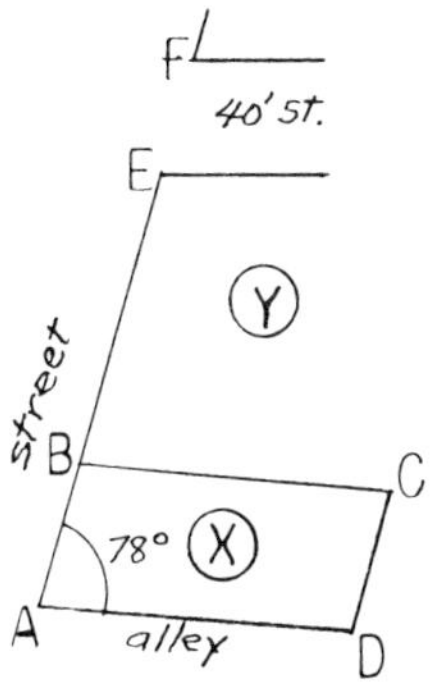

marker (A) at intersection of alley and street, then with street 50.13′ to a stake (B), . . ."

PROBLEM: Locate corners of Lot X.

FOUND: Concrete markers at A, E, F, in good line. 0.3′ north of A is an iron rod. AE measures 150.8′ (surplus of 0.4′); EF measures 40.2′. Concrete markers look undisturbed.

SOLUTION: Concrete marker called for at A has priority over rod nearby; excess of 0.4′ must be prorated between X and Y. Then 50.13 is to x as 150.4 is to 150.8, or x is 50.26′. Measure 50.26′ from A toward EF; found an iron pipe 2″ below ground at 50.20′; bent it 0.06′ toward E. Set transit at A; turn 78° angle from B and find it checks with alley line. Measure plat distance AD: set N/C. Establish C by intersection. Set markers at C and D, if no conflict with owner east of CD.

—MM—

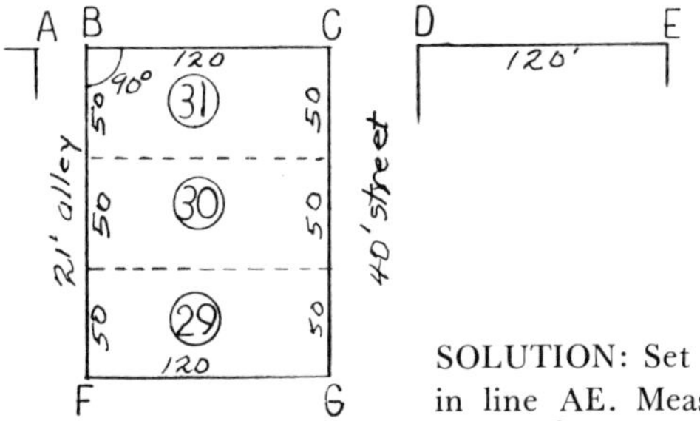

GIVEN: Lots 29, 30, 31 in subdivision.

FOUND: Markers at A, C, E at correct location.

PROBLEM: Set markers at B, F, and G.

SOLUTION: Set transit at C; find it to be in line AE. Measure AB and set marker; turn 90° and measure 150′ and set marker at G. Set F by intersecting 150′ and 120′. Check CF (192.09′).

—NN—

GIVEN: Deed X calls: "Beg. at A, then S 39-34 E 14′ to B, S 48-15 W about 238.93′ to C, N 33-16 W 14′ to D, N 48-15 E about 238.93′ to beg. BPO Y."

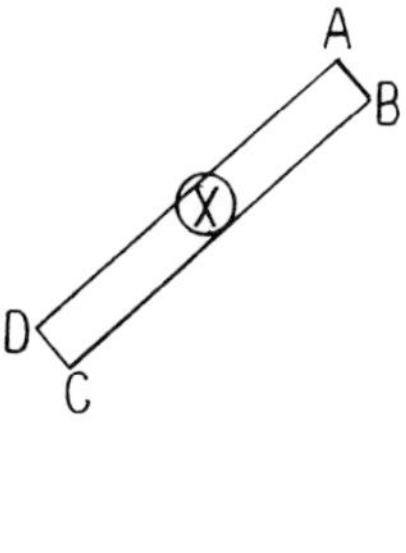

FOUND: Traverse will not close; pipes at C and D are 14.14′ apart.

PROBLEM: Where is error?

SOLUTION: Plat of deeds in area show that bearings are OK and that CD is longer than 14′, and DA slightly less than 238.93′. Scrivener did not figure calls properly.

—OO—

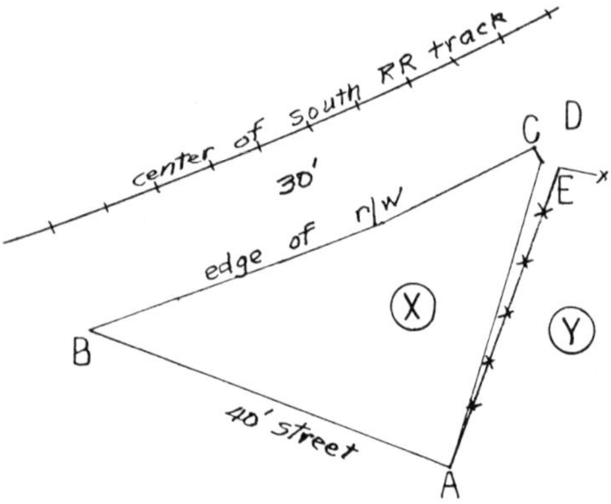

GIVEN: Deed X calls: "Beg. at stake (A) on north edge of street, then N 70 W 104′ to stake at junction of street and south edge of RR r/w (B) then with RR r/w N 71½ E 26′, N 69-45 E 28′, N 68½ E 31′ to stake 30′ from center of RR r/w, N 66 E 44′ to stake, then S 20-45 E 4′ to fence post, then S 20-45 W 89′ to beg." Senior deed for RR r/w says ". . . width to be 60′ . . . an equal width on each side of center line." RR plat shows 2 tracks with south edge of r/w as 30′ south of south track.

FOUND: Iron pipe beside fence post at A about in edge of 40′ uncurbed street. Fence post at E about 35½′ from center of south track. AE measures 89′ and is fenced. Two tracks—westbound and eastbound. Nothing at B, C, D.

PROBLEM: Locate corners B, C, D.

SOLUTION: Intersect 104′ from pipe A with 30′ from center of south track and set H/T at B. Set transit at A, sight on B and turn deed angle toward DE. Measure 89′ toward DE and intersect with 34′ from center of south track and set H/T at D and C. Set transit at B, sight on A and turn deed angle along track; run the 26′, 28′, 31′, and 44′ calls and tie to C. If adjustments are needed at C and D give most weight to RR r/w and distances (bearings could be subject to magnetic attraction). Calculate current calls by bearings from A where there is least attraction. Found fence AE to be off line.

—PP—

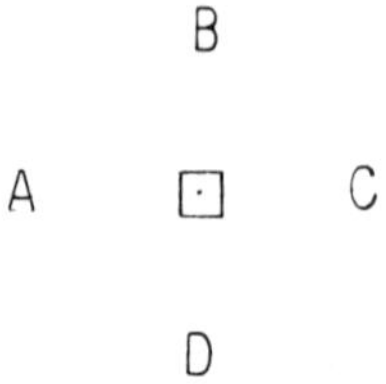

GIVEN: Hub and tack at corner.

PROBLEM: Replace H/T with concrete marker.

SOLUTION: Set steel pin at A and C 12 to 16″ from tack so north edge of pins are in line with center of tack (use plumb bob string as guide). Do same at B and D using east edge of pins. Pry out hub with wrecking bar. Dig hole 10-14″ wide and 24″ deep where hub was (get below frost line). Drive 36″ pipe in center of hole to within ½″ of ground line and bend it to intersection of string lines. Set two 6″ field tile each 12″ long, one on top of other, in hole, fill around them with dirt and tamp well. Fill tile with ready-mix cement. With string adjust pipe to within 0.01′. After cement is partly set, mark date and surveyor's name (use screwdriver imprint). Cost about $2.00. Other ways: Set up 2 transits about 90° to each other and sight on point, or set up, tall tripod and plumb bob over point; hold bob away while setting marker.

—QQ—

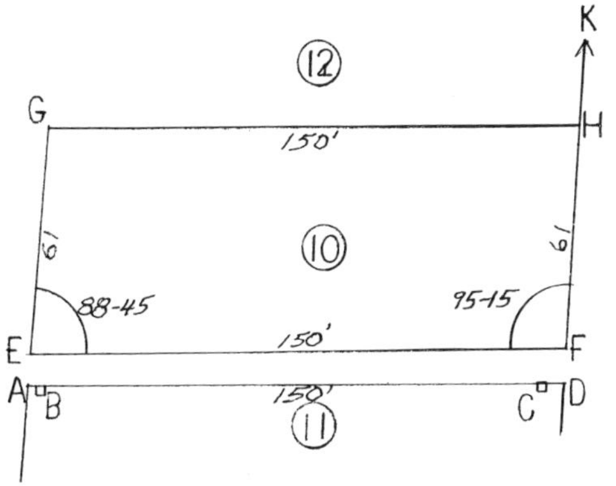

GIVEN: Subdivision plat; no markers.

PROBLEM: Locate corners of Lot 10.

FOUND: 6″ x 6″ concrete markers at B and C, centers of which are 148′6″ apart. IP at K is corner of 4 lots at proper distance from D.

SOLUTION: Extend line BC along north edge of markers and establish A and D 6″ from outside edge of each marker so AD is 150′. Set transit at D; take bearing to A which is N 88-30 W; sight on K and found DK bearing was N 6-45 E which checks with plat. Measure 10.04′ from D toward K (0.04′ is for diagonal) and set H/T at F. Measure 71.04′ toward K and set H/T at H; set transit at A, turn proper angle, measure and set H/Ts at E and G. Check GH for 150′. Set permanent markers.

—RR—

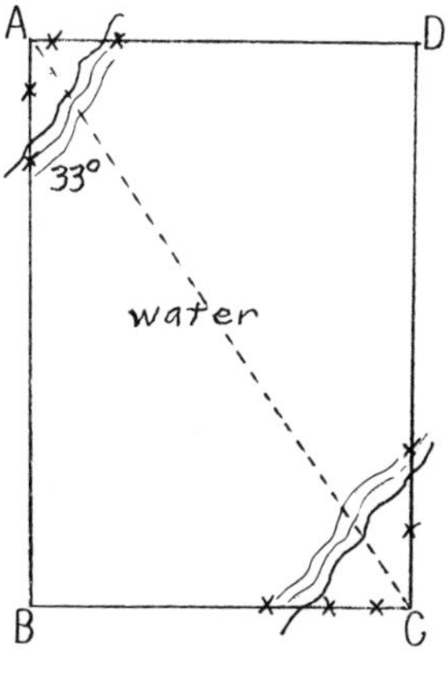

GIVEN: Rectangular parcel ABCD. Perimeter 837.86. All angles 90°. Area 270 acres m or l.

PROBLEM: What is correct area?

FOUND: Water covers all but small area at A and C. C is tree visible from A. Angle BAC is 33°.

SOLUTION: Distance AB + BC = ½ of 837.86 or 418.93. Assume AB is 100, then by formula: a = b tan A or BC = AB x tan 33°, or BC = 100 x 0.6494 = 64.94. And by proportionate measurement 418.93: true AB::164.94:100, or

164.94AB=41893, or AB = 253.99. Then BC = 418.93 - 253.99 = 164.94. Area is 253.99 x 164.94 = 41893.11 square rods or 261.832 acres. This problem contributed by Willis J. Everman, South Shore, Ky.

—SS—

GIVEN: Pipe driven in wrong place.

PROBLEM: Pull pipe and reset.

SOLUTION: With pipe wrench simultaneously rotate pipe and lift up.

—TT—

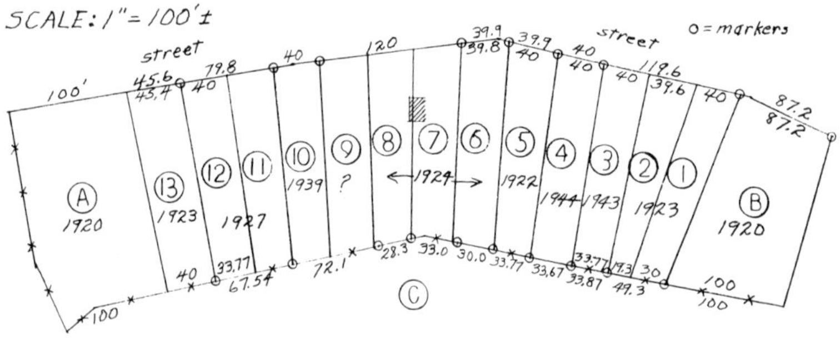

GIVEN: Subdivision plat of 13 lots recorded 1922. Plat lot dimensions shown inside lots; present measurements between markers shown outside lots. Year of lot sale shown. No markers on recorded plat. Small circles are iron pipes found.

PROBLEM: Owners of Lots, 6, 7, and 8 want markers set at rear corners.

FOUND: Subdivision owner's home on Lot 4 built before 1921. Owner of Lot 5 (who bought first lot) helped survey subdivision, set stakes and some pipes, says pipes at rear of Lots 3, 4, 5 are original and correct, but markers at rear of 6, 7, 8 may not be. Front markers check with plat within 0.1′. Rear markers of Lots 6, 7, 8 do not check. A deficit of 10.01′ exists between Lots A and B. Lot 9 at rear has 4.56′ excess. Net rear deficit in Lots 6, 7, 8, 9 is thus 10.01 minus 4.56 or 5.45′. No marker at NW or SW corner of Lot 13. Lots A and B deed distances agree with found distances. No dispute with owner C whose calls neither help nor hurt. Eave of house over line between 7 and 8. Apparent surplus of 5 or 6′ in lot west of A. Fences shown by x's.

CONCLUSIONS: Markers at rear of Lots 1, 2, 3, 4, 5, 11, 12 appear to be original or correct. Surplus west of A cannot be used to offset deficit in subdivision; deficit must be confined to Lots 6, 7, 8, 9, 10 or to Lots 6, 7, 8.

SOLUTION: Obtain concurrence of owners of Lots 6 to 10 to set new rear markers, giving each 1.09′ less than plat shows (this would eliminate encroachment of Lot 7 house eave). Failing in this, set markers at rear of Lots 6, 7, 8 so each has 1/3 of the deficit (1.82′); rear of each lot then is 31.95′ but would not clear dwelling encroachment.

–UU–

For the Reader:

(A) buys 822 acres in 1858.

On 12/9/1873 (A) sells to (B) 200 acres off the SE corner. The north line of the 200 acres begins at a sugar and calls for "west 200 poles to a beech".

On 12/11/1873 (A) sells to (C) 100 acres adjoining part of (B's) north line. (C's) south line begins at the sugar and runs "with (B) west 160 poles to a stake".

In 1892 (C) sells part of his 100 acres to (D), beginning at the same sugar and running "with (B's) line west 80 poles to a lynn, then north xxx".

The sugar, lynn and beech have been found, but the lynn is 11 feet north of a line from the sugar to the beech.

Should (D's) south line run from the sugar to the lynn or from the sugar toward the beech, turning north 11 feet south of the lynn?

INDEX